Interactive
Mathematics Program®

INTEGRATED HIGH SCHOOL MATHEMATICS

FIRST EDITION AUTHORS:
Dan Fendel, Diane Resek, Lynne Alper, and Sherry Fraser

CONTRIBUTORS TO THE SECOND EDITION:
Sherry Fraser, IMP for the 21st Century
Jean Klanica, IMP for the 21st Century
Brian Lawler, California State University San Marcos
Eric Robinson, Ithaca College
Lew Romagnano, Metropolitan State College of Denver
Rick Marks, Sonoma State University
Dan Brutlag, Meaningful Mathematics
Alan Olds, Colorado Writing Project
Mike Bryant, Santa Maria High School
Jeri P. Philbrick, Oxnard High School
Lori Green, Lincoln High School
Matt Bremer, Berkeley High School
Margaret DeArmond, Kern High School District

Key Curriculum Press

Year 1
Second Edition

I M P

This material is based upon work supported by the National Science Foundation under award numbers ESI-9255262, ESI-0137805, and ESI-0627821. Any opinions, findings, and conclusions or recommendations expressed in this publication are those of the authors and do not necessarily reflect the views of the National Science Foundation.

Key Curriculum Press
1150 65th Street
Emeryville, California 94608
email: editorial@keypress.com
www.keypress.com
10 9 8 7 6 5 4 3 2   13 12 11
ISBN 978-1-55953-994-4
Printed in the United States of America

Project Editors
Joan Lewis, Sharon Taylor

Consulting Editor
Mali Apple

Project Administrator
Juliana Tringali

Professional Reviewer
Rick Marks, Sonoma State University

First Edition Teacher Reviewers
Dave Calhoun,
Fresno Unified School District, CA
John Chart, Napa High School, CA
Dwight Fuller, Ponderosa High School, CA
Donna Gaarder,
San Francisco Unified School District, CA
Dan Johnson, Silver Creek High School, CA
Cathie Thompson,
East Bakersfield High School, CA

First Edition Multicultural Reviewers
Edward Castillo, Ph.D.,
Sonoma State University, CA
Genevieve Lau, Ph.D., Skyline College, CA

Copyeditor
Tara Joffe

Interior Designer
Marilyn Perry

Production Editor
Angela Chen

Production Director
Christine Osborne

Production Coordinator
Ann Rothenbuhler

Editorial Production Supervisor
Kristin Ferraioli

Compositor
Lapiz Digital Services

Technical Artists
Greg Reeves, Lapiz Digital Services

Photo Researcher
Laura Murray

Illustrators
Juan Alvarez, Tom Fowler,
Evangelia Philippidis, Diane Varner,
Martha Weston, April Goodman Willy

Cover Designer
Jeff Williams

Printer
RR Donnelley

Executive Editor
Josephine Noah

Publisher
Steven Rasmussen

FOREWORD

Several years ago, I was in your shoes, a ninth grader, starting off my first year of IMP. The "real world" was years away, and the thought of college hadn't even entered my mind. All I knew was that I generally liked school, especially my new math class.

Unlike in my other classes, everyone in my math class participated. The same people who raised their hands in my other classes still raised their hands in my math class, but other people spoke too. I learned that the people who didn't speak up much in other classes (including myself) had a lot to offer! We'd work together to come up with our own ways to solve the problems, which may or may not have been the "textbook" answer. We didn't rely on memorization or on mimicking what we were told. Instead, we were constantly challenged to critically analyze extremely complicated problems, and then, once we figured them out, we had to clearly explain, in writing, our reasoning for why our solution was correct.

Looking back, I appreciate IMP not only for my understanding of the mathematical concepts I learned, but also the communication skills that I developed while tackling those concepts. The ability to persuade people and to effectively argue ideas has proved priceless to me in personal, academic, and professional situations. I encourage you to take advantage of the opportunities that IMP offers to foster your mathematics—and nonmathematics—skills, and most importantly, have fun while doing it!

Kaley Klanica

Kaley Klanica graduated from Eaglecrest High School in Colorado in 1996. Since then, she has received a Bachelor of Arts from Haverford College, a Juris Doctor from Boston University School of Law, and a Master of Public Health from Boston University School of Public Health.

NOTE TO STUDENTS

You are about to begin an adventure in mathematics, an adventure organized around interesting, complex problems. The concepts you learn will grow out of what is needed to solve those problems. Your work will reflect how mathematics is used and the different ways people work and learn together.

Teachers, teacher-educators, and mathematicians have created this book to help you develop the understanding of mathematics and the thinking habits needed for success in this changing world. Over 500,000 students and 2,000 teachers have used these materials since the first edition was published in 1996. This second edition reflects the experiences, reactions, and ideas of students and teachers who have used the first edition for many years.

As you progress through Year 1, you will discover that the problems you work on involve ideas from algebra, geometry, probability, graphing, statistics, and trigonometry. Other mathematical topics will come up later in this four-year program. Rather than studying each of these topics in separate courses, you will find them integrated and presented in meaningful contexts. You will see how they relate to each other and to our world. You will be learning all the essential mathematics that is part of traditional algebra and geometry courses. And you will learn concepts from other branches of mathematics, such as matrix algebra and calculus, as well as probability and statistics.

Each unit in this four-year program has a central problem or theme and focuses on several major mathematical ideas. *A Brief IMP Sampler* provides examples of some of these central problems. As you read the unit descriptions, you will see that the problems involve a variety of contexts. And as you work through the activities, you will come to understand that each unit also involves a variety of mathematical ideas. Supplemental problems at the end of each unit provide opportunities for you to strengthen your understanding and to explore new ideas related to the unit.

Your role as a mathematics student will be as an active learner. You will experiment, investigate, ask questions, make and test conjectures, reflect on your work, and then communicate your ideas and conclusions both orally and in writing. You will do some of your work with your fellow students, just as users of mathematics in the real world often work in teams. At other times, you will work on your own. You will talk about what you are doing and why, and you will present your results to the class.

We hope you enjoy the challenges that the Interactive Mathematics Program presents. And we hope that your experiences give you a deeper appreciation of the meaning and importance of mathematics.

Dan Fendel   Diane Resek

Lynne Alper   Sherry Fraser

CONTENTS

The Game of Pig—Probability and Expected Value

The Overland Trail—Variables, Graphs, Linear Functions, and Equations

The Pit and the Pendulum—Standard Deviation and Curve Fitting

Shadows—Similar Triangles and Proportional Reasoning

A Brief IMP Sampler

You are about to embark on an adventure in the learning of mathematics. One of the features of this program is that you will learn mathematics through the context of problems. You will also see how the mathematics you are learning is used.

Most units have a central problem. You will spend about six to eight weeks developing the concepts and skills needed to solve each of these problems.

Some of these problems are fairly realistic, while others are a bit more fanciful. Here is a sampling of these central problems.

The Game of Pig (Year 1)

This unit centers on a dice game, with the goal of finding the best possible strategy. You will learn basic ideas about probability, as well as more complex ideas such as conditional probability and expected value.

Do Bees Build It Best? (Year 2)

This unit explores the geometry of the honeycomb. The unit problem asks whether the shape that bees use for storing honey is the most efficient. You will extend your understanding of basic ideas such as area and volume, including surface area, and use and prove the Pythagorean theorem.

Cookies (Year 2)

This unit concerns a bakery that is trying to decide how many cookies of each kind to make. The bakery has restrictions on oven space, baking time, and so on. The owners want to use these resources in the way that will maximize profit. You will study linear equations and the graphing of inequalities.

Small World, Isn't It? (Year 3)

This unit looks at world population growth over time and the challenge of making predictions about future growth. You will learn about exponential functions and study rates of change, including derivatives (a fundamental concept in calculus).

High Dive (Years 3 and 4)

This pair of units involves a circus act in which a diver is dropped from a turning Ferris wheel into a moving tub of water. In Year 3, you will use trigonometry to describe circular motion, study the physics of falling objects, and solve the problem using a simplifying assumption. In Year 4, you will figure out how to solve a more realistic but more complex version of the problem. You will use ideas such as vector decomposition to take into account the diver's initial velocity upon leaving the Ferris wheel.

The Pollster's Dilemma (Year 4)

This unit begins with a hypothetical election poll and poses the question of how reliable such a poll might be. You will use ideas about probability and learn about concepts such as sample size, confidence levels, and margin of error.

Patterns

Functions, Reasoning, and Problem Solving

Patterns—Functions, Reasoning, and Problem Solving

The Importance of Patterns

Have you ever noticed how certain things appear in patterns—in regular sequences or forms? Did you know that mathematicians spend a lot of their time looking for *patterns?* They look at the world around them and notice regular sequences and forms, which they then represent with numbers and geometric shapes. That's why this first unit of Year 1 is called *Patterns.*

This unit will also introduce a key mathematical tool for organizing and analyzing number patterns—the In-Out table. This tool is a useful way to work with the concept of **function,** one of the most important ideas in mathematics. You will also be introduced to your first Problem of the Week (POW). Like every activity in IMP, POWs will challenge you not only to *do* the math but also to *understand* the math. You will also get started using powerful technology tools.

Most of all, you will see mathematics is a way of representing and understanding the world you live in, a world full of *patterns.*

Alejandra Pedroza discovers the importance of Patterns.

What's Next?

Mathematics involves looking for patterns: numeric patterns, geometric patterns, all sorts of patterns. In this activity, you will look for patterns in lists.

Each list of pictures or numbers gives the beginning of a **sequence.** Each item on the list is called a **term** of the sequence. Commas separate the terms.

Examine each sequence and look for a pattern. Then write a description of the pattern and a method for finding the next terms of the sequence. Give at least the next three terms.

More than one pattern may fit a given initial sequence. If you see different patterns for a sequence, describe each one.

1.  , , , . . .

2. 1, 1, 2, 1, 3, 1, 4, 1, 5, 1, 6, 1, . . .

3.  , , , , , , , , , . . .

4. 1, 2, 4, . . .

5. 1, 3, 5, 7, 5, 3, 1, 3, . . .

6. 1, 3, 7, 15, 31, . . .

7. Make up a picture sequence of your own. Describe it by giving the first few terms and telling how you would find more terms.

8. Make up a number sequence of your own. Describe it by giving the first few terms and telling how you would find more terms.

Past Experiences

Reflect on your past experiences in learning and using mathematics. With these experiences in mind, write your "mathematics autobiography." Here are some things to think about as you write.

- Consider all the mathematics classes you have taken. What mathematics did you study? What happened in some of those classes that helped you learn? What happened in some classes that made it hard for you to learn?

- Compare experiences you have had working in a group with experiences you have had working alone. When is it best for you to work with others? When is it best for you to work alone?

- Describe a situation when someone helped you learn something difficult.

The Broken Eggs

○ The Situation

A farmer is taking her eggs to market in her cart. Along the way, she hits a pothole, which jars her cart and spills the eggs.

Though the farmer is unhurt, every egg is broken. So she goes to her insurance agent, who asks her how many eggs she had. She doesn't know, but she does remember some things from various ways she tried packing the eggs.

She knows that when she put the eggs in groups of two, there was one egg left over. When she put them in groups of three, there was also one egg left over. The same thing happened when she put them in groups of four, groups of five, or groups of six. But when she put them in groups of seven, she ended up with complete groups of seven, with no eggs left over.

○ Your Task

Your task is to answer the insurance agent's question. In other words,

What can you figure out from this information about how many eggs the farmer had? Is there more than one possibility?

○ Write-up

Your POW write-up will include five components. The components listed below are described in *The Standard POW Write-up.*

1. *Problem Statement*

2. *Process*

3. *Solution*

4. *Extensions*

5. *Self-assessment*

Who's Who?

Steve, Felicia, and Kai—a ninth grader, a tenth grader, and an eleventh grader, but not necessarily in that order—were seated around a circular table, playing a game of Hearts.

Each passed three cards to the person on the right. Felicia passed three hearts to the ninth grader. Steve passed the queen of spades and two diamonds to the person who passed cards to the eleventh grader.

Who was in ninth grade? Tenth grade? Eleventh grade? How were they seated?

1. *Process:* Describe what you did in attempting to solve this problem. Write about such things as

   - how you got started
   - where you got stuck and how you got unstuck
   - how you knew when to stop

   Write about your process of working on the problem, *even if you didn't solve the problem.*

2. *Solution:* Your solution should state what grade each student was in and how they were seated. It should also explain how you can be *absolutely certain* of your answer. Show that your answer fits the information and that it is the *only* answer that fits the information.

From *Mathematics: Problem Solving Through Recreational Mathematics* by Averbach and Chein. Copyright 1980 by W.H. Freeman and Company. Adapted with permission.

The Standard POW Write-up

Problems of the Week (POWs) are an important part of the Interactive Mathematics Program. They will give you experience in carrying out extended investigations of complex problems. They won't always be connected to the rest of the unit.

Despite the name, you will often have more than a week to work on a POW. But you should begin work on these problems as soon as you get them. They take more time than an ordinary homework assignment. You will benefit from working a little bit every night, leaving the POW for a while, and then coming back to it.

POWs give you a chance to write about the mathematics you are doing. You will be expected to explain your thinking more fully in these assignments than in regular class activities and homework. Be sure to leave enough time for this writing.

POWs are accompanied by write-up directions, usually divided into five components. When POW write-ups introduce new components or give specific information, follow those instructions. If a POW lists a component by name only, look back at these descriptions.

1. *Problem Statement:* State the problem clearly in your own words. Your problem statement should be clear enough that someone unfamiliar with the problem could understand what you are being asked to do.

2. *Process:* Describe what you did in attempting to solve the problem. Use your notes as a reminder. Include things that didn't work or that seemed like a waste of time. Do this part of the write-up even

continued ▶

if you didn't solve the problem. If you get assistance of any kind on the problem, tell what the assistance was and how it helped you.

3. *Solution:* State your solution as clearly as you can. Explain how you know that your solution is correct and complete. If you obtained only a partial solution, give that. If you were able to generalize the problem, include your general results. Write your explanation in a way that will be convincing to someone else—even someone who initially disagrees with your answer.

4. *Extensions:* Invent some extensions or variations to the problem. That is, write down some related problems. They can be easier, harder, or about the same level of difficulty as the original problem. (You don't have to solve these additional problems.)

5. *Self-assessment:* Tell what you learned from this problem. Be as specific as you can. Assign yourself a grade for your work on this POW, and explain why you think you deserve that grade.

Inside Out

Supply the missing entries in each In-Out table. If you think there is more than one answer, describe the possibilities. Then write a rule for each table that tells what to do with the *In* to get the *Out*. Express each rule as a complete sentence, such as "The *Out* is 1 more than 4 times the *In*." Be as clear as you can.

1.

In	Out
2	4
3	6
11	22
27	?
?	18

2.

In	Out
0	1
2	7
4	13
7	22
10	31
12	?
?	76

3.

In	Out
house	4
cup	2
writer	5
elephant	7
spin	?
mathematics	?
?	3
?	8
?	0

4.

In	Out
	3
	11
	15
?	7

5.

In	Out
division	I
ever	E
opportunity	O
toast	A
safe	E
people	O
mathematics	?
?	(can't be done)

6. Create two In-Out tables of your own. Give both the table and the rule you used.

Calculator Exploration

Work with a partner on this activity. Your task is to learn whatever you can about how your graphing calculator works. Write down what you discover.

Your discoveries might be as simple as how to do arithmetic operations or as involved as how the calculator works with statistics. You and your partner will determine what to investigate.

Later, you and your partner will be asked to share what you learned and how you learned it. Another pair might have discovered the same thing as you and your partner, but you can still share how you learned it.

Although user manuals may be available, you don't have to use one. The idea is for you to explore the calculator your own way. *Remember to write down everything you discover!*

Pulling Out Rules

1. Write a rule for each In-Out table. Use a complete sentence to describe what to do with the *In* to get the *Out*. Be as clear as you can.

   a. | In | Out |
      |---|---|
      | 10 | 23 |
      | 5 | 13 |
      | 1 | 5 |
      | 0 | 3 |

   b. | In | Out |
      |---|---|
      | 1 | 3 |
      | 3 | 17 |
      | 10 | 66 |
      | 6 | 38 |

   c. | In | Out |
      |---|---|
      | 3 | 17 |
      | 8 | 12 |
      | 15 | 5 |
      | 0 | 20 |

2. These In-Out tables have only one row. Therefore, many rules would fit each table. Find three possible rules for each table. Use a complete sentence to describe how to get the *Out* as a **function** of the *In*.

   a. | In | Out |
      |---|---|
      | 10 | 24 |

   b. | In | Out |
      |---|---|
      | 5 | 25 |

3. This In-Out table gives two rows, which makes it harder to find different rules. Find at least two rules that fit this table.

In	Out
1	2
2	5

4. The supervisor of a community garden project organizes volunteers to help dig out weeds. The more people they have, the more weeds get pulled. The results are better than one might think. Although one person will pull only two bags a day, two people will pull five bags a day, and three people will pull eight bags a day.

   The garden must be cleared of winter weeds. The supervisor estimates that there are 30 bags' worth of weeds to be pulled. How many volunteers are needed to get the job done in a day?

   a. Make an In-Out table that shows the information. Use "Number of people" as the *In* and "Number of bags of weeds pulled" as the *Out*.

   b. Use your In-Out table to solve the problem. Explain your reasoning.

Lonesome Llama

In the land where llamas run free, llamas live in fancy houses decorated with wonderful shapes. Most llamas live in houses that look just like the house of at least one other llama. Llamas who live in identical houses tend to play together. But one llama has a house different from all the rest. Sometimes this llama is left all alone.

The Cards

A set of cards will be distributed, face down, among your group members. Each card has a picture of a llama's house.

One card in the set is a singleton. That is, no other card has a house exactly like it. Every other card has at least one duplicate.

The Task

Your task *as a group* is to discover the singleton card, which shows the lonesome llama's house.

When your group indicates that they have found the lonesome llama's house, the task is ended, whether or not you are correct. Therefore, be sure *everyone* is confident of your answer before you declare that you are done.

The Rules

1. You may *not* show or pass cards to another group member. You may not look at another group member's cards.

2. You may ask questions about the cards that other group members have. You may also make statements or answer questions about your own cards.

3. You may *not* draw pictures or diagrams of the designs on the houses.

4. You may *not* put cards in a common pile.

Aside from these rules, you may work in any way you choose.

Role Reflections

Have you heard the expression, "Two heads are better than one"? When a group thinks about a problem, more brainpower is focused on finding the solution. In groups that work well, everyone takes part. Group members help each other by playing different roles. Read the lists and think of your group, including yourself.

List the task and social roles. Write one group member's name next to a role he or she performed. Describe when that person took that role. There may be roles that no one performed, at least this time around.

Task Roles

- *Getting Started:* Helping the group get organized to work together
- *Seeking Information:* Asking for facts, suggestions, or ideas
- *Giving Information:* Offering facts, suggestions, or ideas
- *Clarifying:* Clearing up confusion by asking questions, suggesting new approaches, or explaining ideas so everyone understands
- *Finding Agreement:* Asking questions or making statements that test to see if the group is in agreement and a conclusion has been reached

Socio-Emotional Roles

- *Encouraging:* Being friendly and responsive to others, accepting others and their contributions, listening, showing regard for others
- *Expressing Group Feelings:* Sensing feeling, mood, or relationships within the group; sharing one's own feelings with others
- *Harmonizing:* Attempting to reconcile disagreements, reducing tension, getting people to explore their differences
- *Compromising:* Offering to compromise one's own position, ideas, or status; admitting error
- *Gatekeeping:* Keeping the whole group involved in discussion
- *Setting Standards:* Expressing standards to help the group achieve their goals

Adapted from *Group Processes in the Classroom* by Patricia Smuck and Richard A. Smuck. (Dubuque, IA: William C. Brown, 2005).

Communicating About Mathematics

You learned how to organize number information using In-Out tables. Now, you will explore how these tables can be used to understand other real-world and mathematical situations.

You will read the story of a woman named Marcella, who shares her bagels with everyone she meets. And you will finish your first POW *(The Broken Eggs)* and start a new one *(Checkerboard Squares)*. Using In-Out tables to organize the data in all of these problems will help you solve them.

How many ways do you think you can combine the numbers 1-2-3-4 to make a different number? When you do these number combinations with a calculator, you will realize how lucky you are to have such a tool for handling the rules for order of arithmetic operations.

Have you ever lived where the temperature dipped below zero degrees—or "into the negative numbers," as the weather reporter on TV sometimes says? By spending some time "cooking" with some chefs who use what they call "hot and cold cubes," you will understand more about positive and negative numbers.

Tenley McGurk describes her group's explanation for computing with positive and negative numbers.

Marcella's Bagels

Have you ever been really in the mood to eat a bagel? Some pretty amazing things can get in the way of this pursuit.

Marcella was walking home from the bakery near the beach one day. She had just bought a big bag of bagels to share with her daughter, Sonya. Sonya loves bagels.

On the side of the road, two people were collecting food for families in need. Marcella decided that she had quite a few bagels in her bag. Sonya didn't need *that* many bagels.

"Here," said Marcella. "You can have half of my bagels." The people were very happy to get all of those bagels. Marcella thought for a moment and then said, "Aw, take one for each of yourselves." So there went two more.

As Marcella walked along the beach, some surfers came out of the water. They saw, and even smelled, the fresh bagels. "Could you, by any chance, spare a few bagels?" they pleaded. "We are so-o-o hungry after riding all of those gnarly waves."

As you might imagine, Marcella was not thrilled. But she had a good heart and recognized hunger when she saw it, so she handed her bag to the surfers. They took half of her bagels. Then, just as they were about to hand the bag back, they took two more.

continued ▶

Now, Marcella was a very reasonable person who liked to help others. She thought she still had enough bagels left to make Sonya happy. She walked on.

As you may already have guessed, Marcella didn't get far before she had another encounter. Just before she reached home, her friend Susan approached. After saying hello, Susan explained that she was on her way to get some bagels for her family. She seemed to be in a bit of a rush.

Generosity overtook Marcella. She found herself saying, "Why don't you save yourself the trip and take some of my bagels? As you can see, I've got several." So Susan took half of what Marcella had in the bag and then two more.

Marcella finally reached her front door. When she opened her once bulging bag of bagels, she discovered that there were only two left! She had a simple bagel lunch with her daughter Sonya, and then there were none.

After lunch, Sonya asked her mother how many bagels had been in the bag to begin with. Marcella told her the story of her walk. She then said that if Sonya could figure it out herself, Marcella would take her rollerblading in the park the next day.

Sonya took a while, but she eventually figured it out and got her rollerblading outing. What was Sonya's answer?

1-2-3-4 Puzzle

You can change the meaning of an arithmetic **expression** by inserting or removing parentheses. Another way to change the meaning of an expression is to rearrange its terms.

In this activity, you will use the digits 1, 2, 3, and 4, in any order, to create arithmetic expressions with different numeric values.

For this problem, a "1-2-3-4 expression" is any expression that uses each of these digits *exactly once,* according to these rules.

- You may use any of the four basic arithmetic operations—addition, subtraction, multiplication, and division. For example, $2 + 1 \cdot 3 - 4$ is a 1-2-3-4 expression for the number 1.

- You may use **exponents.** For example, $2^3 - 4 - 1$ is a 1-2-3-4 expression for the number 3.

- You may juxtapose two or more digits—that is, put them next to each other—to form a number such as 12. For instance, $43 - 12$ is a 1-2-3-4 expression for the number 31.

- You may use **square roots.** For example, $\sqrt{4 \cdot 2 + 1}$ is equal to 3, so $3 + \sqrt{4 \cdot 2 + 1}$ is a 1-2-3-4 expression for the number 6.

- You may use **factorials.** For example, 4! means $4 \cdot 3 \cdot 2 \cdot 1$, so $3 + 4! + 1 - 2$ is a 1-2-3-4 expression for the number 26.

- You may use parentheses and brackets. For example, $1 + 4 \cdot 3^2$ is a 1-2-3-4 expression for the number 37. You can add parentheses and brackets to get $[(1 + 4) \cdot 3]^2$, which is a 1-2-3-4 expression for the number 225.

Your task is to create a 1-2-3-4 expression for each of the numbers from 1 through 25. *Remember:* Every expression must use each of the digits 1, 2, 3, and 4 *exactly once.*

Uncertain Answers

Mathematical expressions are simplified according to this **order of operations.**

  1 Parentheses

  2 Exponents

  3 Multiplication and division (equal priority) from left to right

  4 Addition and subtraction (equal priority) from left to right

1. *Fix these equations.* None of these statements is correct as written. Rewrite each, inserting parentheses in the expressions on the left so that the resulting statements are correct **equations.** Check your results with your calculator.

   a. $12 - 8 \cdot 2 + 7 = 36$

   b. $8 - 15 + 6 \div 3 = 1$

   c. $7 + 3^2 = 100$

   d. $24 + 16 \div 8 - 4 = 10$

   e. $20 \div 7 - 2 + 5^2 \cdot 3 = 79$

2. *What could it be?* Place parentheses in different places in these expressions to see how many different values you can make for each expression. Use your calculator to confirm your results. Find at least three values for each problem.

   a. $7 - 5 \cdot 8 + 6 \div 2$

   b. $4 + 9 - 6 \div 2 \cdot 5 + 1$

   c. $4 - 3 - 2 + 1$

3. Create two more "1-2-3-4 expressions" to add to your class collection.

Extended Bagels

The extensions section of the POW write-up is often an important part of your written report. It gives you the opportunity to examine what the critical elements of the problem are and how they could be changed.

In this activity, you will work on an extension to *Marcella's Bagels*. Your work should give you some ideas about how to use an In-Out table to gain insight into a problem situation.

Here's the extension.

> *How does the solution to* Marcella's Bagels *depend on the number of bagels Marcella has when she gets home?*

To explore this extension, try different values for the number Marcella has when she gets home, like 3, 4, 5, or even 0. Assume that everything in the problem is the same as in the original problem except for this number.

Make an In-Out table of your results. The *In* will be the number of bagels Marcella has when she gets home. The *Out* will be the number she must have started with. (Your answer from *Marcella's Bagels* will be the *Out* for the case where the *In* is 2.)

Once you have several entries in your table, look for a relationship between the *Out* and the *In*. Then find a rule to describe this relationship. You may find it helpful to look for a pattern in the *Out* column.

The Chefs' Hot and Cold Cubes

You may have learned some rules for doing arithmetic with positive and negative numbers. Many people find these rules hard to remember and don't understand where they come from.

This story will help you understand how positive and negative numbers work. Many people remember the story many years after they first heard it. The memory helps them reconstruct the rules for positive and negative numbers. The story also helps make sense of the rules.

The Story

In a far-off place, there was once a team of amazing chefs who cooked up the most marvelous food ever imagined.

They prepared their meals over a huge cauldron, and their work was delicate and complex. They frequently had to change the temperature of the cauldron to bring out the flavors and cook the food to perfection.

They adjusted the temperature by adding either special hot cubes or special cold cubes to the cauldron or by removing some of the hot or cold cubes that were already in the cauldron.

The cold cubes were similar to ice cubes, except they didn't melt. The hot cubes were similar to charcoal briquettes, except they didn't lose their heat.

If the number of cold cubes in the cauldron was the same as the number of hot cubes, the temperature of the cauldron was 0 degrees on their temperature scale.

continued ▶

For each hot cube put into the cauldron, the temperature went up 1 degree. For each hot cube removed from the cauldron, the temperature went down 1 degree. Cold cubes worked the opposite way. Each cold cube put in lowered the temperature 1 degree. Each cold cube removed raised it 1 degree.

The chefs used positive and negative numbers to keep track of the temperature changes.

For example, suppose 4 hot cubes and 10 cold cubes were dumped into the cauldron. The temperature would be lowered by 6 degrees altogether. This is because 4 of the 10 cold cubes would balance out the 4 hot cubes, leaving 6 cold cubes to lower the temperature 6 degrees. To represent these actions and their overall result, the chefs would write

$$^+4 + {}^-10 = {}^-6$$

Similarly, if they added 3 hot cubes and then removed 2 cold cubes, the combined result would raise the temperature 5 degrees. In that case, the chefs would write

$$^+3 - {}^-2 = {}^+5$$

The subtraction indicates that the 2 cold cubes were being *taken out*.

If they wrote $^-5 - {}^+6 = {}^-11$, it would mean that 5 cold cubes were added and then 6 hot cubes were removed. The combined result was to lower the temperature 11 degrees.

Sometimes the chefs wanted to raise or lower the temperature by a large amount, but they did not want to put the cubes into the cauldron one at a time. So, they would put in or take out bunches of cubes.

For instance, to raise the temperature 100 degrees, the chefs might toss 5 bunches of 20 hot cubes each into the cauldron, instead of 100 cubes one at a time. This saved a lot of time because they could have assistant chefs do the bunching.

When the chefs used bunches of cubes to change the temperature, they used a multiplication sign to record their activity. For example, to describe tossing 5 bunches of 20 hot cubes each into the cauldron, they would write

$$^+5 \cdot {}^+20 = {}^+100$$

continued ▶

The chefs could also change the temperature by removing bunches. For example, suppose they removed 3 bunches of 5 hot cubes each. This would lower the temperature 15 degrees, because each time a bunch of 5 hot cubes was removed, the temperature went down 5 degrees. To record this change, they would write

$$^-3 \cdot {}^+5 = {}^-15$$

The $^-3$ means that three bunches were being removed. The $^+5$ shows that there were 5 hot cubes in each bunch.

1. Each problem describes an action by the chefs. Figure out how the temperature would change overall in each situation. Write an equation to describe the action and the overall result.

   a. Three cold cubes were added, and then 5 hot cubes were added.

   b. Five hot cubes were added, and then 4 cold cubes were removed.

   c. Two bunches of 6 cold cubes each were added.

   d. Four bunches of 7 hot cubes each were removed.

   e. Three bunches of 6 cold cubes each were removed.

2. Describe the action involving hot or cold cubes that is represented by each arithmetic expression. State how the temperature would change overall.

   a. $^+4 - {}^-3$

   b. $^-6 + {}^-4$

   c. $^-10 \cdot {}^-5$

   d. $^+4 \cdot {}^-8$

Do It the Chefs' Way

Explain each expression in terms of the "hot and cold cubes" model. Your explanations should describe the action and state how the temperature changes overall.

1. $^-6 + {}^-9$
2. $^-7 - {}^-10$
3. $^+5 \cdot {}^-2$
4. $^-4 - {}^+6$
5. $^+3 + {}^-7$
6. $^-6 \cdot {}^+9$
7. $^-3 \cdot {}^-4$
8. $^+8 - {}^-12$
9. $^-12 + {}^+5$

Checkerboard Squares

This is a standard 8-by-8 checkerboard made up of 64 small **squares.** These squares can be combined to form squares of other sizes within the checkerboard.

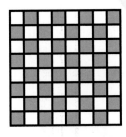

For example, look at the 3-by-3 square outlined within the larger checkerboard. This is just one of many 3-by-3 squares you could find in this board.

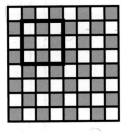

1. How many squares of various sizes are on an 8-by-8 checkerboard altogether?

When you are confident that you have counted *all* the squares on an 8-by-8 checkerboard, move on to the generalization in Question 2.

2. Suppose you have a square checkerboard of some other size (not 8-by-8). How can you determine how many squares are on it altogether?

   You will know you are done when no matter what size checkerboard you are given, you can give a clear procedure for easily computing the total number of squares.

○ Write-up

1. *Problem Statement*

2. *Process:* Be sure to describe any diagrams or materials you used.

3. *Solution:* Explain your answer in the case of the 8-by-8 checkerboard in complete detail. Also give results, with explanations, for any other checkerboard sizes you investigated. Include your reasoning when you write about any generalizations you found.

4. *Extensions*

5. *Self-assessment*

You're the Chef

After a lengthy reign as the "Number One Chef in the World," you have decided to step down and retire to the warmer climates of the Lazy Chef Sunset Ranch. It has been a memorable time, but one with many responsibilities.

Though it was fun to be a master chef, any drastic miscalculation of the temperature could have caused catastrophic results.

Your final responsibility before retirement is to train an assistant chef to take your place. Assistant chefs spend most of their time putting hot and cold cubes into bunches. They know very little about changing the temperature in the cauldron. But they do know how to do arithmetic with whole numbers.

Prepare a manual for the assistant chef who will be taking your place. Include specific examples of all the different ways to raise and lower the temperature of the cauldron. Also describe how the chefs would write your examples as arithmetic expressions.

Investigations

Over the next several days, you will be working on a problem called *Consecutive Sums*. You and your group will probably not be able to learn all there is to know about consecutive sums in just a few days. But you can look at many specific examples, and you will probably see some general principles.

When a skateboarder says he's done a "180" or a "360," he is describing how far he has spun around in the air, using degrees as his measurement—just like mathematicians measure the degrees of an angle. You will measure the angles in closed geometric figures, called **polygons,** and make some **conjectures** about what you discover. These will lead to you expressing your conclusions in algebraic **formulas.**

Conjectures are educated judgments that you make after you have gathered some, but not all, of the information about something. What do you think is true about a given situation? What are you certain is true, and what makes you certain? These questions underlie the idea of **proof**—one of the big ideas in mathematics.

A group of students uses pattern blocks to measure angles.

Consecutive Sums

A sequence of two or more whole numbers is **consecutive** if each number is one more than the previous number.

For example, the numbers 2, 3, and 4 are consecutive; the numbers 8, 9, 10, and 11 are consecutive; and the numbers 23 and 24 are consecutive.

On the other hand, the numbers 6, 8, and 10 are not consecutive, because each number is *two* more than the previous number. A single number by itself is not considered consecutive.

A **consecutive sum** is a sum of consecutive numbers. Each of these expressions is a consecutive sum.

$$2 + 3 + 4$$
$$8 + 9 + 10 + 11$$
$$23 + 24$$

These examples illustrate ways to express 9, 38, and 47 as consecutive sums.

For this activity, consider only consecutive sums involving positive **whole numbers** (1, 2, 3, 4, and so on). These are also called the **natural numbers** or **counting numbers.**

Your task is to explore consecutive sums. Look for patterns and make generalizations.

A good strategy is to look at specific cases. For instance, you might try to find all the ways to write each number from 1 to 35 as a consecutive sum. Are there numbers for which this is easy to do? Are there any that are impossible?

continued ▶

Your group will produce a display on a large sheet of paper. It will show your results and include summary statements of the patterns you found. Your group will also present some of its discoveries and strategies to the class.

Include conjectures (general patterns that you think might be true) as well as patterns that you are certain are true. Include any explanations you find of why your patterns are true.

Also, if you made a conjecture and later discovered that it was false, include both the original conjecture and the evidence that convinced you it was false.

Add It Up

Summation notation can be useful when working with sums of numbers, such as consecutive sums. For instance, we can express the consecutive sum $10 + 11 + 12$ as

$$\sum_{r=10}^{12} r$$

This expression is read, "The summation, from r equals 10 to 12, of r." The symbol Σ is an uppercase letter in the Greek alphabet called *sigma*.

The expression $\displaystyle\sum_{i=2}^{6} i$ means $2 + 3 + 4 + 5 + 6$. This could also be written as $\displaystyle\sum_{n=2}^{6} n$. It doesn't matter which letter is used.

This **sigma notation** can be used for sums more complex than sums of consecutive numbers. For example, $\displaystyle\sum_{t=5}^{8}(4t^2 + 3)$ represents the expression

$$(4 \cdot 5^2 + 3) + (4 \cdot 6^2 + 3) + (4 \cdot 7^2 + 3) + (4 \cdot 8^2 + 3)$$

In the expression $\displaystyle\sum_{t=5}^{8}(4t^2 + 3)$, the number 5 is called the *lower limit,* the number 8 is called the *upper limit,* and the expression $4t^2 + 3$ is called the *summand.*

1. Write each summation as a string of numbers or numeric expressions added together.

   a. $\displaystyle\sum_{z=3}^{8} z$ b. $\displaystyle\sum_{m=1}^{5} 2m$ c. $\displaystyle\sum_{c=2}^{9}(4c + 7)$

continued ▸

2. Use summation notation to describe the number of squares in this diagram.

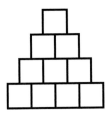

3. Use summation notation to express each sum.

   a. 10 + 11 + 12 + 13 + 14 + 15

   b. 3 + 6 + 9 + 12 + 15 + 18 + 21

   c. 8 + 11 + 14 + 17 + 20

4. Use summation notation to describe the total number of small squares in this diagram.

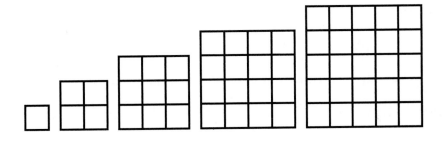

Group Reflection

As you are learning, people play many roles when they work in groups. Of course, this is true not only in math classes.

This activity is an opportunity for you to reflect upon the way you participate in groups. Be as thoughtful as possible when you answer these questions. Consider your work during the past few days on *Consecutive Sums,* as well as in previous classroom activities.

1. Try to remember a time when you or someone in your group was left out of a discussion.

   a. Describe the situation. Did anyone try to include that person? If not, why not? If yes, how?

   b. What might you have done to help with the situation?

2. Think of a time when someone made a mistake in your group.

   a. What did you think or say?

   b. How do you think groups should handle mistakes by group members?

3. Remember a time when you thought of saying something, or did not understand something, but were afraid to speak out.

   a. Describe the situation, what you wanted to say, and why you didn't say it.

   b. How do you wish you had handled the situation?

4. Do you participate more or less than other group members? Why do you think you do so?

5. Discuss how the amount of homework preparation you do for class affects your participation in group discussions.

That's Odd!

A student who worked on the consecutive sum problem made this conjecture, based on many examples.

> If an odd number is greater than 1, then it can be written as the sum of two consecutive numbers.

If you think this statement is false, your task is to find a **counterexample.** A counterexample to a statement is an example showing that the statement is false. A counterexample to this statement would be an odd number greater than 1 that *cannot* be written as the sum of two consecutive numbers.

If you think this statement is true, your task is to create a set of general instructions for writing any odd number greater than 1 as the sum of two consecutive numbers. Illustrate those instructions using a specific example. Your instructions should work for *any* odd number greater than 1.

Pattern Block Investigations

Part I: Pattern Block Designs

1. Work together as a group to create a single group design from your blocks. Your group should have access to at least a half tub's worth of pattern blocks.

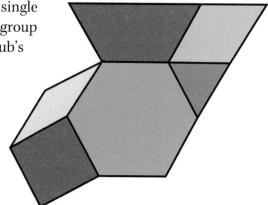

2. Explore different ways to fit several pattern blocks together so they come together at a single point. Can you do this using blocks that are all the same shape? If so, which shapes can you do this with?

Part II: Pattern Block Angles

You have seen that you can think of an **angle** as "an amount of turn." You can express the size of an angle as a fraction of a complete turn. For example, if you are facing north and turn to face west, you have made a quarter turn. A turn of this amount is called a **right angle.**

You can also measure angles in terms of **degrees,** where a complete turn is defined to be 360 degrees (written 360°); so, 1 degree is $\frac{1}{360}$ of a complete turn. Because a right angle is one-fourth of a complete turn, it is measured as 90°.

An angle between 0° and 90° is called an **acute angle,** and an angle between 90° and 180° is called an **obtuse angle.**

Mathematicians formally define an angle to be a figure formed by two **rays** with a common initial point, such as $\angle BAC$.

To measure such an angle, you can imagine that you are standing at the **vertex** A, facing toward B. You want to know how much you would need to turn to face toward C.

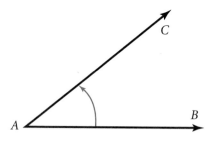

continued ▶

The angles in a polygon are the angles formed at each vertex by two of the polygon's sides.

3. Use the fact that a complete turn is 360° to find the measure (in degrees) for each angle of each of the six different pattern blocks. Use the blocks to explain your reasoning.

Part III: Pattern Block Angles with a Protractor

A protractor is a tool used to measure amount of turn, just like a ruler is a tool to measure length. For example, two rays drawn to indicate one-quarter of a complete turn will measure 90° on your protractor.

4. Now that you know the measures of each pattern block angle, learn how to use the protractor to find this measurement. To begin, trace the pattern blocks on paper and extend the sides so the angle you wish to measure is more evident.

Degree Discovery

In the activity *Pattern Block Investigations,* you found the angles of some very special polygons.

In this activity, you will use protractors to explore more general polygons and to look at the sum of the angles.

1. Begin with **triangles.** Use a ruler or straightedge to draw a variety of triangles. Measure the angles for each triangle. Then find the sum of the angles for each triangle.

2. What **conclusion** do your results suggest? Does this conclusion hold for the angles you found for the triangle pattern block?

3. Now do the same for **quadrilaterals.** Does your conclusion hold for the angles you found for the various quadrilateral pattern blocks?

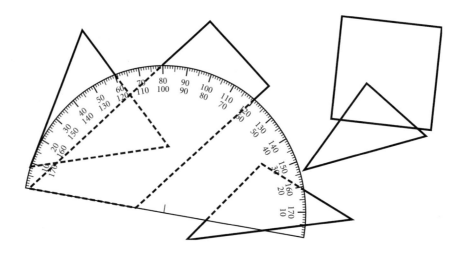

Polygon Angles

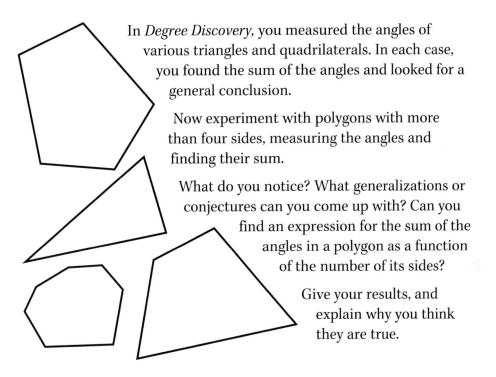

In *Degree Discovery,* you measured the angles of various triangles and quadrilaterals. In each case, you found the sum of the angles and looked for a general conclusion.

Now experiment with polygons with more than four sides, measuring the angles and finding their sum.

What do you notice? What generalizations or conjectures can you come up with? Can you find an expression for the sum of the angles in a polygon as a function of the number of its sides?

Give your results, and explain why you think they are true.

An Angular Summary

Through your work and class discussion on *Polygon Angles,* you have seen a formula that gives the sum of the angles in a polygon in terms of the number of sides it has.

1. Summarize what you know about the sum of the angles in a polygon. Explain the reasoning behind any formulas you include.

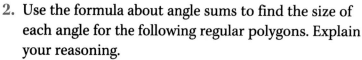

A **regular polygon** is a polygon that has all of its angles equal in size and all of its sides the same length.

2. Use the formula about angle sums to find the size of each angle for the following regular polygons. Explain your reasoning.

   a. A regular **pentagon** (5 sides)

   b. A regular **octagon** (8 sides)

3. Draw each polygon in Question 2, using a protractor to get the angles to the right size. You can decide on the lengths for the sides in each case. The sides of your pentagon do not need to have the same length as the sides of your octagon.

Putting It Together

This unit lays a foundation for everything that follows in IMP. You have learned to look for patterns, to organize and make conjectures about what you find, and to put your discoveries into numbers and writing. You have worked both with other students and on your own.

The final days of *Patterns* are a chance to put it all together. You will finish your work on *POW 2: Checkerboard Squares*. You will apply ideas you have learned in the unit to new problems that involve both numbers and geometry. As you search for the patterns in all of these problems, you may also begin to see how algebra—including **variables**—can help you extend the mathematics you are learning in IMP.

The final activity in *Patterns* lets you show what you have learned so far. You will finish the unit by putting together a portfolio—a record of your work. The portfolio will give you the chance to solidify your thinking about the unit's mathematics.

Joe Gonzales presents his results about an In-Out table.

Squares and Scoops

1. Squares are stacked in piles of different heights.

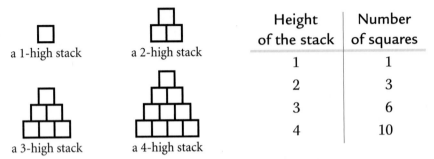

a 1-high stack   a 2-high stack

a 3-high stack   a 4-high stack

Height of the stack	Number of squares
1	1
2	3
3	6
4	10

   a. Use diagrams or a continuation of the table to find the number of squares in a 7-high stack.

   b. How many squares are in a 10-high stack?

   c. How many are in a 40-high stack?

   d. Give a general description for how to find the number of squares in an *n*-high stack.

2. Explore the different ways you can arrange scoops of ice cream on a cone. Each scoop is a different flavor. The pictures show one scoop and two scoops. The table gives more information.

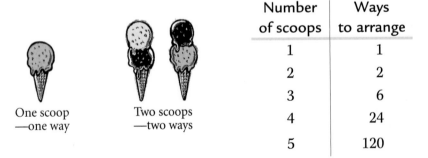

One scoop
—one way

Two scoops
—two ways

Number of scoops	Ways to arrange
1	1
2	2
3	6
4	24
5	120

   a. Show why the *Out* for three scoops is equal to 6.

   b. Find a numeric pattern for the entries in the table. Use that pattern to find the number of ways to arrange 7 scoops and the number of ways to arrange 10 scoops.

   c. Describe how you would find the number of ways to arrange 100 scoops. You don't need to find this number. Just describe how you would find it.

Another In-Outer

Find the missing items in each In-Out table.

For Questions 1, 4, and 5, give a description in words for how to find the *Out* from the *In*.

For Questions 2, 3, and 6, give an algebraic equation for the *Out* as a function of the *In*.

1.

In	Out
●	LBC
○	SWC
■	LBS
△	SWT
□	?
?	LWT

2.

In	Out
2	−6
5	−15
0	0
13	?
−7	?
?	−30
?	35

3.

In	Out
1	5
3	11
7	23
10	32
−2	−4
−5	?
?	2
?	−19

×3 +2

4.

In	Out
Ruth	U
Johnny	I
Carol	S
Anne	O
Aaron	S
Robert	?

5.

In	Out
	6
	11
	19
	?

Hair² + #eyes

6.

In	Out
1	−2
4	−11
−5	16
0	1
3	?
−2	?
?	10

x−3 + 1

Diagonally Speaking

How many diagonals does a polygon have?

A **diagonal** is a **line segment** that connects two vertices of a polygon but is not a side of the polygon. For example, in this polygon, segment *AD* is a diagonal, and segment *CD* is a side.

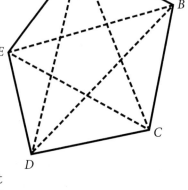

The diagonals are shown as dashed lines. This polygon has five diagonals.

As you might expect, the number of diagonals in a polygon depends on the number of sides the polygon has.

1. Experiment by drawing polygons and finding out how many diagonals each has. Organize the results in an In-Out table. The *In* will be the number of sides of the polygon. The *Out* will be the number of diagonals.

2. Look for a pattern in your table.

   a. Describe that pattern.

   b. Once you have found a pattern, use it to figure out how many diagonals a 12-sided polygon has. Try to confirm your result by actually counting the diagonals.

3. Explain why your pattern holds true. That is, why should the number of diagonals in a polygon follow your pattern?

The Garden Border

Leslie was planning an ornamental garden.

She wanted her garden to be square with 10 feet on a side, including a tile border. She planned to use part of this area for a border of tiles. The square tiles were each 1 foot by 1 foot.

Leslie had to figure out how many tiles she needed.

Your challenge is to figure out how many tiles Leslie needed, without counting the tiles individually. Write down as many ways as you can for doing this. Describe in detail the arithmetic involved.

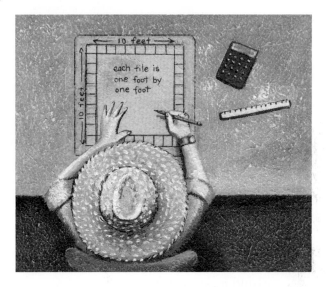

For each method you find, draw a diagram that indicates how that method works.

Border Varieties

Leslie decided it would be nice to have a general formula for her border problem. The formula would give the number of tiles needed as a function of the garden's size.

She imagined a square garden that was *s* feet on each side. The tiles would be squares that measure 1 foot on each side.

She asked for some help from students who had worked on the border problem for the 10-by-10 square in *The Garden Border*. Because they had solved that problem in different ways, they also came up with different formulas for the general problem.

For example, one student counted 10 tiles along each edge and then subtracted 4, because the corner tiles had each been counted twice. This student's arithmetic looked like this.

$$
\begin{array}{r}
10 \\
\times\ 4 \\
\hline
40 \\
-\ 4 \\
\hline
36
\end{array}
$$

continued ▶

Using this diagram to explain the arithmetic, this student came up with the formula $4s - 4$ for the general border problem.

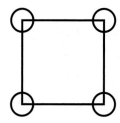

Each problem below shows the arithmetic used by another student for the 10-by-10 case, along with a diagram that explains the arithmetic. For each method, do two things.

• Explain how the arithmetic matches the diagram.

• Find a general formula that fits that student's way of thinking about the problem. Your formula should use s to represent the length of one side of the garden. Make your formula match the arithmetic as closely as possible.

1.

$$\begin{array}{r} 10 \\ + 10 \\ \hline 20 \end{array} \qquad \begin{array}{r} 8 \\ + 8 \\ \hline 16 \end{array}$$

$$\begin{array}{r} 20 \\ + 16 \\ \hline 36 \end{array}$$

2.

$$\begin{array}{r} 10 \\ 9 \\ 9 \\ + 8 \\ \hline 36 \end{array}$$

3.

$$\begin{array}{r} 8 \\ \times 4 \\ \hline 32 \\ + 4 \\ \hline 36 \end{array}$$

4.

$$\begin{array}{r} 100 \\ - 64 \\ \hline 36 \end{array}$$

5.

$$\begin{array}{r} 9 \\ \times 4 \\ \hline 36 \end{array}$$

Patterns Portfolio

To complete your portfolio for *Patterns*, you will do three things.

• Write a cover letter in which you summarize the mathematics in the unit.

• Select and organize papers from among your work in the unit.

• Write a reflection that discusses your personal growth during the unit.

Cover Letter

Look back over *Patterns*. Write a cover letter describing the unit's central ideas. This description should include a summary of key mathematical concepts and ideas about what mathematics is and how mathematics is learned.

As part of your portfolio, you will select specific activities on certain topics. Your cover letter should include an explanation of why you select the items you do.

Compiling Papers

Look through your papers for *Patterns*. Select the items described here for your portfolio. Later you'll add the in-class and take-home assessments for *Patterns*.

• *Past Experiences*

Include this so you can look back to compare your experiences before this year with your experiences this year and beyond.

• At least two activities on In-Out tables

Choose activities that helped you understand what In-Out tables are and how to use them.

continued ▶

- One or two activities focusing on proof

  Choose activities in which you explained clearly how you knew your answer was correct and complete.

- *You're the Chef*

  You may want to refer to this summary of the hot and cold cube model in later work with positive and negative numbers.

- Two additional examples of your mathematical writing, including at least one Problem of the Week (either *POW 1: The Broken Eggs* or *POW 2: Checkerboard Squares*)

Personal Growth

Write about your personal development during this unit. You may want to address these questions.

- How do you feel you progressed
  - in working with others?
  - in presenting to the class?
  - in writing about and describing your thought processes?
- What do you feel you need to work on? How might you work on it?

Include any other thoughts about your experience with this unit that you want to share with a reader of your portfolio.

Supplemental Activities

The supplemental activities for each unit explore some of the themes and ideas that are important in that unit.

Here are some examples of the supplemental activities for *Patterns.*

- *Whose Dog Is That?* and *Infinite Proof* help continue your work with proof.

- *Three in a Row* and *The General Theory of Consecutive Sums* follow up on your investigation of consecutive sums.

- *Diagonals Illuminated* gives two ways to extend your thinking from *Diagonally Speaking.*

Other supplemental activities for *Patterns* follow up on other assignments and continue your work with In-Out tables.

Keep It Going

1. Examine the first several terms in each sequence. Look for a pattern that explains how the sequence is formed. Then write a description of the pattern and a method for finding the next few terms. Give at least the next three terms.

   a. 1, 4, 9, 16, . . .

   b. 2, 6, 18, 54, . . .

   c. 1, 4, 7, 10, . . .

   d. 1, 3, 6, 10, . . .

2. Find the 100th term for the sequences in Questions 1a, 1c, and 1d. Explain how to get these values without calculating all of the previous terms.

The Number Magician

A magician chooses a volunteer from the audience and says, "Pick a number, but don't tell me what it is. Add 15 to it. Multiply your answer by 3. Subtract 9. Divide by 3. Subtract 8. Now tell me your answer."

"Thirty-two," replies the volunteer.

The magician *immediately* guesses the volunteer's number.

1. What was the volunteer's number?

2. The magician couldn't possibly have worked backward that fast. How did the magician find the answer so quickly?

From *Mathematics: Problem Solving Through Recreational Mathematics* by Averbach and Chein. Copyright 1980 by W.H. Freeman and Company. Adapted with permission.

Whose Dog Is That?

Abigail, Beth, Celia, Dayna, and Eudora bring their dogs to the Little Red Schoolhouse for Dogs. During the morning break, they find that each of their pets has the same name as the husband of one of the other women.

In particular, they notice these facts.

- Abigail's dog is named George.
- Celia's dog is named Jerry.
- Eudora's dog is named Ike.
- Dayna's dog is named Frank.
- Abigail's husband is also named Frank.
- Beth's husband has the same name as George's wife's dog.
- Horace and his wife, Celia, have the best-behaved dog.

Your job is to figure out the names of each woman's husband and each woman's dog.

In addition, you must explain clearly how you found your answer. If you think there is more than one possibility, give all solutions, and explain why there are no more. If you think there is only one solution, explain why there are no others.

From *Mathematics: Problem Solving Through Recreational Mathematics* by Averbach and Chein. Copyright 1980 by W.H. Freeman and Company. Adapted with permission.

A Fractional Life

Here is a problem that is part of the *Greek Anthology*, a group of problems collected by ancient Greek mathematicians.

> Demochares has lived a fourth of his life as a boy, a fifth as a youth, a third as a man, and has spent 13 years in his dotage.

How old is Demochares?

Note: The phrase "in his dotage" refers to the period of Demochares' old age.

Adapted from Problem 6.13 from *An Introduction to the History of Mathematics*, 5th ed., by Howard Eves. (New York: Holt, Rinehart and Winston, Inc., 1983). Reproduced by permission of the publisher.

Counting Llama Houses

Here is one of the houses from *Lonesome Llama.*

The llama houses all have the same basic form, but they vary in their details.

1. Describe the ways in which the houses differ.

2. How many different houses could be created using the variations found on the *Lonesome Llama* cards? Explain your reasoning.

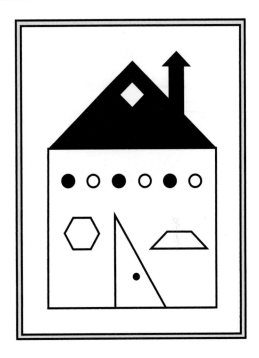

It's All Gone

A man goes into a store and says, "If you give me as much money as I have with me now, I will spend $10 in your store." The store owner agrees, and the man spends the money.

He goes into a second store and again says, "If you give me as much money as I have with me now, I will spend $10 in your store." Again, the owner agrees, and the man spends the money.

In a third store, he repeats his proposition. The proprietor agrees, and the man spends the money.

At this point, the man has no money left.

How much money did he have to begin with? Explain your answer.

1-2-3-4 Varieties

In *1-2-3-4 Puzzle,* you were asked to express each number from 1 through 25 as a 1-2-3-4 expression.

As you may recall, a 1-2-3-4 expression is an arithmetic expression that uses each of the digits 1, 2, 3, and 4 exactly once, according to certain rules.

This new problem involves one additional rule.

The digits 1, 2, 3, and 4 must appear in numerical order.

For example, you can express the number 10 as $1 + 2 + 3 + 4$ or as $1 \cdot 2 \cdot 3 + 4$, but not as $4 \cdot 3 - 2 \cdot 1$ or as $3 + 2 + 1 + 4$.

Can the numbers from 1 through 25 all be written as 1-2-3-4 expressions using this new rule? Support your answer with examples.

You might also explore these questions.

- What is the largest result you can get from a 1-2-3-4 expression using the extra rule?
- What other variations can you create for *1-2-3-4 Puzzle?*

Positive and Negative Ideas

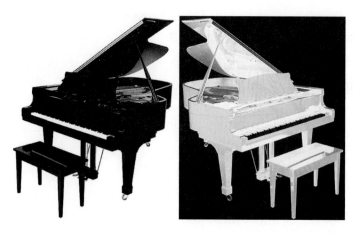

The Chefs' Hot and Cold Cubes presents one way of thinking about positive and negative **integers** and how to do arithmetic with them.

How do *you* like to think about these numbers? What mental or physical models do you use to understand arithmetic with integers?

Describe at least one other model for working with integers. Give examples of how to explain the arithmetic operations using your model.

Chef Divisions

The introductory discussion of the hot and cold cubes model left out one of the four basic arithmetic operations—division.

What do you suppose the chefs mean by the expression $^+15 \div {}^-3$? What might the numbers $^+15$ and $^-3$ in this division problem represent in terms of the "hot and cold cubes" model?

1. Explain $^+15 \div {}^-3$ in terms of the model.

2. Make up some other division problems using integers. Explain them using the model. Include different combinations of signs.

3. Suppose a and b represent any integers. Give a general description of what an expression of the form $a \div b$ means. What does a tell you? What does b tell you?

More Broken Eggs

In *POW 1: The Broken Eggs,* you found a possible number of eggs the farmer might have had when her cart was knocked over.

You may have found only one solution to that problem, but there are actually many solutions.

Your task now is to look for other solutions to the problem. Find as many as you can. If possible, find and describe a pattern for getting all the solutions. Also explain why all the solutions fit that pattern.

Here are the facts you need to know.

- When the farmer put the eggs in groups of two, there was one egg left over.

- When she put them in groups of three, there was also one egg left over. The same thing happened when she put them in groups of four, five, or six.

- When she put the eggs in groups of seven, she ended up with complete groups of seven, with no eggs left over.

Three in a Row

In *That's Odd!,* you looked at the conjecture that any odd number greater than 1 can be written as a sum of two consecutive numbers. Now you will look at sums of more than two consecutive numbers.

1. Consider the case of sums of three consecutive numbers.

    a. Develop a conjecture for how to describe the numbers that can be written as such sums.

    b. Once you come up with a conjecture for part a, try to prove it. Your proof will show that all numbers that can be written as such a sum fit your description.

    c. Try to formulate and prove a conjecture that goes the other direction. Show that all numbers that fit your description can be written as such a sum.

2. Look for a generalization that holds for sums of other lengths. Try to prove it.

    Suggestion: Start by looking at sums for another odd number of terms. You may want to consider consecutive sums of integers and not restrict yourself to natural numbers.

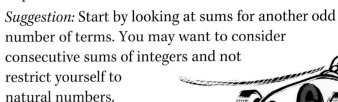

Any Old Sum

This problem, like *Consecutive Sums,* is about sums of natural numbers. That is, it's about sums of whole numbers other than zero. But this time you aren't restricted to consecutive sums.

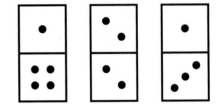

Now your task is to look at *all* the ways various numbers can be written as a sum of natural numbers. For example, the number 4 can be written as a sum in exactly eight ways.

$$1 + 1 + 1 + 1 \qquad 1 + 2 + 1$$
$$2 + 2 \qquad\qquad\quad 4$$
$$2 + 1 + 1 \qquad\quad 3 + 1$$
$$1 + 3 \qquad\qquad\quad 1 + 1 + 2$$

Notice that 4 by itself is counted as one way. Also, $1 + 2 + 1$, $2 + 1 + 1$, and $1 + 1 + 2$ are counted separately.

Explore. Look for patterns. Examine methods of categorizing the different ways to write a number as a sum. Make some generalizations.

State each generalization clearly. Explain why each generalization is always true.

Getting Involved

Imagine that you have been in the same group for about a week. During that time, everyone in your group has been doing fine except one person.

This person doesn't say a thing besides "I don't know" and won't help with presentations.

Write about

- why someone might behave that way
- what you and the group can do about it

Look for many possibilities, and be specific.

The General Theory of Consecutive Sums

In *Consecutive Sums,* you were restricted to positive whole numbers (1, 2, 3, 4, and so on). The idea of consecutive sums also makes sense for all integers. For example, $-2 + -1 + 0 + 1$ is a consecutive sum for the number -2.

Your task is to investigate how your results on *Consecutive Sums* would have been different if you had used the complete set of integers—positive, negative, and zero. You can use the hot and cold cubes model to help with the arithmetic of the sums.

In particular, you may want to look for a general rule for the number of ways in which a given integer can be expressed as a consecutive sum.

If you worked on *Three in a Row,* you may find your results from that activity helpful.

Start with many examples. Look for patterns in your data. Then think about ways in which you might explain those patterns.

Infinite Proof

Proof involves considering all the possible cases.

For example, in *Who's Who?*, you saw that there is a **unique** solution to the problem. One way to prove this is to examine each possible combination. Because the problem has only a finite number of combinations, you can consider each combination individually. You would then see that only one of them fits the problem.

This activity involves proving things about every possible case in situations with infinitely many cases to consider.

1. You know that there are infinitely many odd numbers. Prove that the **square** of *every* odd number is odd.

2. A **prime number** is a number greater than 1 whose only whole-number **divisors** are 1 and itself. For example, 3 and 7 are prime numbers. But 12 is not a prime number, because it has whole-number divisors other than 1 and 12. For example, 4 is a divisor of 12.

    Prove that *every* prime number greater than 10 must have the digit 1, 3, 7, or 9 in the 1s column.

Different Kinds of Checkerboards

In *POW 2: Checkerboard Squares,* you found a way to compute the total number of squares that can be formed on an *n*-by-*n* checkerboard.

What if the checkerboard isn't necessarily square? For example, how many squares are there all together on a 4-by-6 checkerboard?

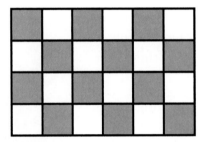

Here are two squares that can be formed on a 4-by-6 checkerboard.

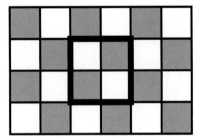 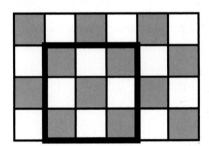

Start with this 4-by-6 example, and then look for ways to generalize what you find.

Your goal is to find a method to compute the total number of squares that can be formed on an *m*-by-*n* checkerboard.

Lots of Squares

Can you divide a square into a certain number of smaller squares? The answer may depend on exactly how many smaller squares you want.

The first diagram shows that any square can be divided into four smaller squares. The second diagram shows that any square can be divided into seven smaller squares.

1	2
3	4

Notice that these smaller squares don't have to be the same size. However, the smaller portions must all be squares, not simply **rectangles.**

In this activity, you will investigate what numbers of smaller squares are possible. For example, you can probably see that there is no way to divide a square into just two smaller squares. (Try it and convince yourself that it is impossible.)

1. Start with specific cases. Is it possible to divide a square into three smaller squares? Five? Six? Eight? (The cases of four and seven are shown in the diagrams, although you may want to look for other ways to do them.)

   Continue this process, at least up to the case of 13 smaller squares.

2. Now reflect on what you've done. Imagine continuing this process. Would there be any numbers beyond 13 for which you *couldn't* divide a square into that many smaller squares? What patterns can you find in the cases you've done that help with this question?

3. What is the largest "impossible" case? Prove your answer. That is, prove that all cases beyond the one you named are possible.

A Protracted Engagement

Deon and Marsha thought of an interesting way to announce their long-awaited wedding.

They decided to send out the invitations in code. In the code they chose, each letter is represented by an angle of a certain size.

An angle between 0° and 5° represents the letter *A*, an angle between 5° and 10° represents *B*, an angle between 10° and 15° represents *C*, and so on.

To avoid confusion, they never used angles that were exact multiples of 5°. The design below spells out part of their invitation.

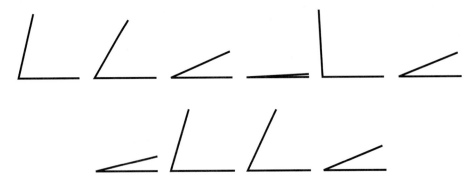

1. What does this message say?

2. What would be the biggest angle that might occur in Deon and Marsha's code?

3. Make up a message of your own that is between 10 and 20 letters long. You and other students can exchange and decipher one another's messages.

A Proof Gone Bad

Jerry thought he had it all figured out. He was supposed to be writing a proof to show that the sum of the angles of a quadrilateral is 360°, based on the fact that the sum for any triangle is 180°. But when he looked over his work, it seemed to show that the sum for a quadrilateral is 720°.

Here's what he wrote.

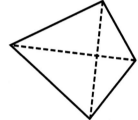

> Take any quadrilateral, such as the figure formed by the solid lines.
>
> Then draw its diagonals (shown as dotted lines). As you can see, this breaks the quadrilateral into four small triangles. We are assuming that the angles of each triangle add up to 180°. The sum of the angles of the quadrilateral is 4 · 180°, which is 720°. Because this can be done with any quadrilateral, it must be true that the sum of the angles of any quadrilateral is 720°.

1. Explain what went wrong with Jerry's proof.

2. Write a correct proof based on Jerry's diagram.

From Another Angle

As you saw in *An Angular Summary,* a regular polygon is a polygon in which all the angles are the same size and the sides are the same length.

For example, these three pattern blocks are regular polygons.

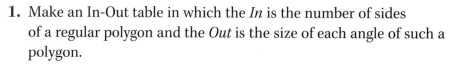

The problems in this activity involve both pattern blocks and regular polygons. In doing these problems, you may use any formulas you know for the sum of the angles in a polygon.

1. Make an In-Out table in which the *In* is the number of sides of a regular polygon and the *Out* is the size of each angle of such a polygon.

   For example, if a regular polygon has three sides, each angle of that polygon is 60°, because the three angles must add up to 180°.

   Look for a rule for your table.

2. Investigate whether any other regular polygons, other than those shown in Question 1, can be made from pattern blocks.

   a. For each polygon that can be made, draw a picture to show how it can be made.

   b. Explain why your answer to part a is complete. That is, explain why no other regular polygons can be made from pattern blocks.

Careful! One polygon that can be made has more than six sides.

You may want to use pattern blocks to do this assignment. For your convenience, the six pattern blocks are shown here drawn to scale.

From One to N

In Question 1 of *Squares and Scoops*, you looked at the number of squares in an n-high stack.

You may have seen that this number could be found by getting the sum $1 + 2 + \cdots + n$.

For example, the number of squares in a 5-high stack is $1 + 2 + 3 + 4 + 5$. The sum $1 + 2 + \cdots + n$ occurs often in mathematics problems.

Your task in this activity is to find a simple expression in terms of n that allows you to find this sum without repeated addition. (What you are looking for is called a *closed formula*.)

If you find such an expression, look for a proof that your answer is correct. Don't just say, "It works." You need to guarantee that it works for *every* value of n.

Diagonals Illuminated

In *Diagonally Speaking,* you looked at a pattern for finding the number of diagonals in a polygon. Your goal was to find a way to get the number of diagonals as a function of the number of sides of the polygon. This activity will help you gain more insight into diagonals.

This simple problem shows two ways to think about In-Out tables and functions.

A group of people are in a room. How many hands do they have?

A table for this problem might begin like the one shown at right.

In	Out
1	2
2	4
3	6
4	8

In the table, you might notice that the *Out* values go up by 2 at each step. You can explain this pattern by the fact that when each new person joins the group, two more hands are added. Using this idea, each *Out* value is found from the previous *Out* value. Such a method for getting the *Out* values is called a **recursive formula.** To use this approach, you have to list the *In* values in order, going up by 1 at each step.

You might also notice that each *Out* value is twice the corresponding *In* value. In other words, the table can be described by the formula $Out = 2 \cdot In$. You can explain this by the fact that each person has two hands. In this approach, the *Out* is expressed directly in terms of the *In*. The equation $Out = 2 \cdot In$ is called a **closed formula.** With this approach, you can list the *In* values in any order you like.

continued ▶

Part I: A Diagonal Closed Formula

In *Diagonally Speaking,* you may have found a recursive formula. In other words, you described how the number of diagonals grows as the number of sides increases by one side at a time. But a recursive formula can be difficult to work with for polygons that have many sides. For example, you probably wouldn't want to use a recursive formula to find the number of diagonals for a 1000-side polygon.

Your task in Part I is to find a closed formula for the diagonal problem. That is, look for a formula that will allow you to find the number of diagonals for any polygon directly in terms of the number of sides, without having to work your way through all the polygons that have fewer sides.

Part II: A Diagonal Explanation

Whatever method you discovered for finding the number of diagonals in Part I, you probably found it by looking at some examples and seeing a pattern.

That's an excellent approach, but it doesn't necessarily tell you why the pattern holds. If you aren't sure why a pattern holds, you don't have much of a guarantee that it will always hold.

Your task in Part II is to explain why your method for finding the number of diagonals must work. Whether you are using a recursive formula or a closed formula, you will need to think about what a diagonal is and not just look at the numeric pattern in your data.

More About Borders

The problem in *Border Varieties* involved finding the number of tiles needed to form a border around an *s*-by-*s* square. In this activity, you will explore some variations or extensions of that problem.

1. Suppose Leslie's garden isn't square. For example, if she has a garden that is 6 feet by 8 feet, the tiles would look like this diagram.

   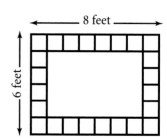

   Explore examples like this. Then develop an expression for the number of tiles needed for the border of a garden that is *m* feet by *n* feet.

2. Suppose that each tile costs $3 and that Leslie buys topsoil for the part of the garden that is not tiled. It costs 20 ¢ for each square foot of ground to be covered with topsoil. Develop an expression, in terms of *m* and *n*, for Leslie's total cost for tiles and topsoil for a garden that is *m* feet by *n* feet.

3. Consider the problem of creating a border 2 feet wide. For example, for a garden 10 feet by 10 feet, the border would look like the diagram. How many tiles would be needed? In general, how many tiles would be needed for a border like this for a square garden that is *s* feet by s feet? What about for a rectangular garden that is *m* feet by *n* feet? What would Leslie's costs be, based on the information in Question 2?

   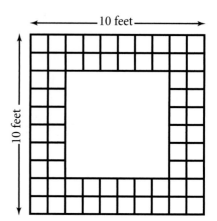

4. Generalize the problem even further by considering a border that is *r* feet wide all around.

Programming Borders

In *More About Borders,* you generalized the ideas in the original border problem.

The generalization would allow someone running an outdoor supply store to answer questions about different border designs or the amount of topsoil needed.

Technology can help you use the generalizations. Write a program that answers some or all of the questions posed in *More About Borders.*

At the ultimate level of detail, the program might run something like this display. The program does everything but the numbers that follow the question marks. The user enters the numbers 7, 10, and 2.

> How wide is
> the garden?
> ? 7
> How long is
> the garden?
> ? 10
> How wide is
> the border?
> ? 2
> You will need 52 tiles
> and will have to cover
> 18 square feet with
> topsoil.
> This will cost
> 159.60 dollars.

POW-Style Write-up of "Marcella's Bagels"

Problem Statement

Marcella is carrying a bag of bagels on her way home. Various people stop her three times along the way, each time taking *half of her bagels plus two more*. By the time she gets home, she has only two bagels left! Sonya gets a trip to the park if she can figure out how many bagels Marcella had to start with.

Process

When I got home, I started to tell the story to my little brother. All he wanted to know was where the bagels were. He was no help!

I asked myself what if Marcella had 100 bagels to start with? The first group would have taken 50 and then 2 more, leaving her with 48. The second group would have taken half of that, 24, and 2 more, leaving her now with 22. The third person would have taken 11 and 2, and Marcella would have 9 left. That's too many!

At this point, I noticed the jar of kidney beans on the kitchen table. I spilled a bunch out on the table. My dad is used to my using beans for my math homework.

I decided it might help to work backward. I used 2 beans to represent the 2 bagels that Marcella had left at the end of her walk.

continued ▶

Because the third person, Susan, had taken half plus 2 more, I added 2 more bagels to my pile to represent the 2 extra Susan had taken. Now I had 4.

This was only half of what Marcella had before Susan came along. So I doubled the 4, giving me 8. That's how many Marcella had when she met Susan, so she also had this many after she left the group of surfers.

The surfers had taken half of what she had plus 2 more. So I put those 2 back into Marcella's pile, now giving her 10. This was what she had after the surfers had taken half.

So she must have had 20 bagels before they came along.

continued ▶

The group collecting for families in need had also taken half plus 2 more, so I added the 2 more. This gave Marcella 22.

Finally, I doubled that.

This gave her 44 bagels.

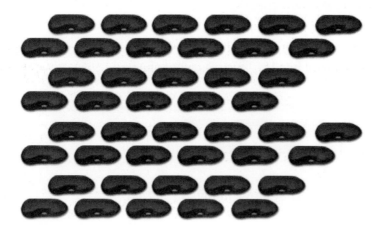

Solution

Because each bean represents a bagel, Marcella must have started with 44 bagels.

At this point, my brother asked why I was playing with beans again. So I told him the story of Marcella and the bagels a second time. This time, I acted it out with the beans as I talked.

$$44 \to 22 \to 20 \to 10 \to 8 \to 4 \to 2$$

The first two steps ($44 \to 22 \to 20$) are what happened when Marcella met the people collecting for those in need, and so on.

continued ▶

It really helped to use beans as Marcella's bagels. I knew I had the correct answer, because when I acted it out, I came out with 2, which is what she had left. I don't know if another answer would work, but I don't think so because I worked backward and didn't have any choices along the way. It wasn't like Marcella had fewer than 5 bagels or more than 1 bagel. She had *exactly* 2 bagels left.

The pictures I drew could now help me explain my process to someone else.

Extensions

What if Marcella had 5 bagels left? What would that mean about how many she had at the start? What if she ended with 2 but each group she met took a third and then 3 more? What if she met even more people on her way home?

Self-assessment

I learned a couple of things. First, I learned about experimenting and trying things out. That really helped me understand the problem when I made a guess like 100 bagels. I also learned about working backward, because that's how I got my final answer. And I learned that using physical objects can help in understanding a problem and in explaining it to someone else.

I would give myself an A– for this work. I described how I worked on the problem, I found the answer, and I have given a very good explanation. To get an A, I probably should have tried to answer some of my extensions, but I didn't have time.

The Game of Pig

Probability and Expected Value

The Game of Pig—Probability and Expected Value

Chance and Strategy

The theory of **probability** was developed in the seventeenth century, primarily to answer questions posed by gamblers. Mathematicians of the time saw that certain strategies could help increase the chances of success in many types of situations, even in those involving luck. Today, probability is used in many areas, from scientific research to helping people make good business decisions.

You'll enter the world of probability by exploring a dice game called Pig. In the first few activities of this unit, you will begin your search for the best **strategy** for the game by experimenting with different options.

Flipping coins is another way to investigate ideas about probability. In *The Gambler's Fallacy,* you'll use coin flips to arrive at a conclusion that may surprise you. Then you'll move on to the formal definition of *probability.* As you'll see, people study probability through both experimentation and theoretical analysis.

Leah Allen and Crystal Kovarik begin the unit by playing Pig and thinking about strategies they might use to play the game.

The Game of Pig

Pig is a dice game. To play, you need an ordinary six-sided die, labeled 1 to 6.

Each turn of the game consists of one or more rolls of the die. You continue to roll until you decide to stop or until you roll 1. You may choose to stop rolling at any time.

Scoring

If you choose to stop rolling *before* you roll 1, your score for that turn is the sum of all the numbers you rolled on that turn.

If you roll 1, your turn is over, and your score for that turn is 0.

Each turn is scored separately.

Examples

- You roll 4, 5, and 2 and decide to stop. Your score for this turn is 11.
- You roll 3, 4, 6, and 1. The turn is over because you rolled 1. Your score for this turn is 0.

Your Assignment

1. Play several turns as a whole group. As you play, talk about how you are deciding whether to roll again.

2. Split into two smaller groups and play some more.

3. List some of the strategies that you used. Be prepared to discuss them with the class.

A Sticky Gum Problem

This Problem of the Week, or POW, starts with some specific situations. It then asks you to generalize what you've learned from the situations.

1. Ms. Hernandez comes across a gumball machine one day when she is out with her twins. Of course, the twins each want a gumball. They also insist on having gumballs of the same color. They don't care what color the gumballs are, as long as they're both the same.

   Ms. Hernandez can see that there are only white gumballs and red gumballs in the machine. The gumballs cost a penny each, and there is no way to tell which color will come out next. Ms. Hernandez decides to keep putting in pennies until she gets two gumballs of the same color.

   Why is three cents the most she might have to spend?

2. The next day, Ms. Hernandez and her twins pass another gumball machine. This one has three colors of gumballs: red, white, and blue.

   What is the most Ms. Hernandez might have to spend at this machine to get matching gumballs for her twins?

3. Mr. Hodges and his triplets pass the three-color gumball machine. Of course, his children insist that they all get the same color gumball. What is the most Mr. Hodges might have to spend?

After you have answered these questions, create some examples of your own. You may want to begin with more examples about the Hernandez twins, using different numbers of colors. Or you may want to create examples using the three-color gumball machine and larger sets of children.

continued ▶

As you create and solve examples of your own, look for a way to organize the information. Also look for patterns. Your goal is to find a formula that will guarantee each child a gumball of the same color. If someone tells you the number of gumball colors and the number of children, your formula will tell you the maximum amount of money the parent might need to spend.

○ *Write-up*

Begin your write-up for this POW with a discussion of Questions 1 through 3. Explain your answers to each question, and describe the process you used to solve them.

Then discuss the problems you made up and their solutions. Explain how you organized your information and the patterns you found.

Finally, state any general ideas you were able to formulate. Include conjectures you may have about the general problem, even if you can't prove them. For each general statement, explain why you think it's true. Provide examples to illustrate each statement. Describe the process by which you arrived at that generalization.

Adapted from "A Sticky Gum Problem" in *aha! Insight* by Martin Gardner (New York: W.H. Freeman and Company, 1978).

Pig at Home

Play Pig with someone at home. If you don't have a die, you can put numbers on a wooden cube. Or you can write the numbers 1 through 6 on cards and put them in a bag. If you use cards, mix the cards well each time before drawing a card. Replace the card after each draw.

Write about your experience playing Pig. Include the following in your write-up:

- Who did you play with?
- How did you teach them to play?
- What strategies did each of you use, and which strategy seemed most effective?
- Did you play *with* one another or *against* one another? Why?

Pig Strategies

Continue working in your group to develop strategies for playing Pig.

1. Share within your group the strategies you each used in *Pig at Home.*

2. Discuss the various strategies. Decide which strategy is the best that you have found so far. It may be one of those written for *Pig at Home,* or it may be something new.

3. Write a clear explanation of this best strategy so that another group would be able to use it to play Pig.

Waiting for a Double

In many games that use dice, such as backgammon, you roll two dice at a time. Special rules often apply when you roll a double. (A *double* means the same number is showing on both dice.)

When you play games like these, you might want to know how long it takes to get a double. Here is an experiment to consider.

You roll a pair of dice and continue rolling until you get a double. You record the number of rolls it took to get a double.

Example:

First roll	Dice show 3 and 4.
Second roll	Dice show 2 and 5.
Third roll	Dice show 5 and 3.
Fourth roll	Dice show 2 and 2.

In this example, it took four rolls to get a double.

Now answer these questions.

1. Predict the **average** number of rolls it will take to get a double. Write a sentence or two explaining why you made that prediction.

2. Do the experiment ten times. That is, for each experiment, roll a pair of dice until you get a double, counting how many rolls it takes. Write down the number of rolls needed each time.

3. Use the data you gathered in Question 2 to answer these questions.

   a. What was the greatest number of rolls it took to get a double? What was the least?

   b. What was the average of the ten experiments?

4. How close is the average you found in Question 3b to the prediction you made in Question 1? Would you revise your prediction now, based on your experiment? Why or why not?

The Gambler's Fallacy

In the game of roulette, a ball is spun around a roulette wheel, and it lands in either a red slot or a black slot. There is also a very small chance that it will land in a green slot, but in this problem, we will simplify things by ignoring that fact.

The chance of the ball landing in a red slot is the same as its chance of landing in a black slot.

Some gamblers use the following strategy for playing roulette: If the ball lands in red a certain number of times in a row, they bet that black will be next, because they figure it is black's turn. Similarly, if a string of blacks occurs, they bet on red, because they figure red will be more likely than black after a string of blacks.

A Historical Example

In a famous incident in 1913, at the Casino in Monte Carlo, black came up a record 26 times in a row. By about the fifteenth time, people started betting overwhelmingly on red, believing that it was now red's turn. As a result, the Casino made an enormous sum of money.

The Experiment

Do this experiment with a partner.

> Flip a coin 25 times. Record each flip as heads (H) or tails (T), according to the outcome.

When you have completed all 25 flips, you will have a list of 25 letters, made up of H's and T's.

Starting from the beginning of your list, find the first instance of three identical flips in a row (either three heads or three tails). We will call three identical flips in a row a *triplet*.

continued ▶

Record whether the flip that followed this first triplet was the *same* as the letters in the triplet or *different* from them. Then move to the next triplet. Again, record whether the flip that followed it was the same as or different from the letters in the triplet. Continue in this way through your whole list. If there's a triplet at the end of your 25 flips, ignore it, because nothing follows it.

Finally, count how many "sames" and how many "differents" you got. That is, if you have four identical flips in a row, then that is two triplets. For example, suppose you have H H H H T as part of your record. The first three H's form a triplet that is followed by an H, which is the *same* as the triplet. The second, third, and fourth H's also form a triplet, followed by T, which is *different* from the triplet.

The first triplet is followed by H, which is the same as the triplet.

The second triplet is followed by T, which is different from the triplet.

Similarly, if you have five identical flips and then a different flip (such as T T T T T H), that gives three triplets. The first two triplets are followed by a flip the *same* as the letters in that triplet. A *different* letter follows the third triplet.

Expecting the Unexpected

Suppose you flip a coin a bunch of times. Even though heads and tails are equally likely, you know that you still might get more heads than tails or more tails than heads.

Experiment with the following and think about what is likely.

1. Gather some data by flipping a coin 50 times and recording the number of heads you get. Then do this 50-flip experiment again and record the results.

2. Each of your classmates will also do this 50-flip experiment twice, recording the number of heads each time.

   a. What do you think will be the greatest number of heads that anyone will get in a 50-flip experiment? The least number of heads?

   b. About how many of the experiments do you think will result in exactly 25 heads?

Coincidence or Causation?

Most people accept the idea that a coin's "history" doesn't affect the probability of a given result in the future. But there are times when previous events *do* affect future events.

In each situation, decide whether you think the past will or will not have a certain influence on the future. In each case, state your decision and write a paragraph explaining why you believe it is correct.

1. A baseball player has averaged hitting a home run once every seven games this season. He has just hit a home run in each of the previous three games.

   Are his chances of hitting a home run in the next game greater than, less than, or no different from usual?

2. It seems to Mr. Bryant that every time he comes along Pine Street to the traffic light at Lincoln Avenue, the light is red. He is so infuriated that he contacts the city planner. The city planner reports that the light is set so that cars on Pine Street and cars on Lincoln Avenue are given equal time to pass through the intersection.

   If you are driving right behind Mr. Bryant one morning and come to that traffic light, do you think that your chances of getting a red light are greater than, less than, or equal to those given by the city planner?

3. The Happy Days Ice Cream Cone Company claims that, on average, only about 1 out of every 100 boxes of their famous ice cream cones will contain a broken cone. The company gladly replaces any box containing a broken cone.

   You purchase a box of Happy Days Ice Cream Cones. Upon arriving home, you discover that one of the cones is broken. You decide to return the box and exchange it for a new box. Just to be sure, you immediately check the new box for broken cones.

   Is the chance that the new box contains a broken cone different from 1 out of 100?

What Are the Chances?

Part I: Finding Probabilities

Sometimes the only way to find the probability of something is to use **observed probability**—a model based on using your own experience or making an educated guess. In other cases, you can use **theoretical probability.**

Items A through I pose questions about probability. In each case,

- Decide on the probability, using a theoretical model if you can.
- Describe how you decided on the probability. State whether your answer was based on a theoretical model or on observed results, or whether it was just a pure guess.

A. You pull one gumball out of a bag that contains three red gumballs, two blue gumballs, and four black gumballs. What is the probability that the gumball you picked is blue?

B. What is the probability of snow falling in Florida at least once next July?

C. You arrive at an intersection with a traffic signal. What is the probability that the light is red?

continued ▶

D. You flip a coin twice. What is the probability of getting one head and one tail?

E. You roll a standard die. What is the probability of getting a prime number?

F. Your teacher selects two students at **random** from your class to run an errand. What is the probability that you are one of the two students?

G. You randomly point to a student in your mathematics class. What is the probability that this student is wearing sneakers?

H. You roll two dice. What is the probability of getting doubles?

I. You roll a pair of dice until you get doubles. What is the probability that you get doubles in three or fewer rolls?

Part II: Probabilities on the Number Line

Make a number line like the one below. For each item in Part 1, indicate its probability by putting the letter in the proper place on the number line.

0   1

Paula's Pizza

Paula's favorite pizza place offers six toppings: sausage, onions, mushrooms, pepperoni, olives, and peppers.

Paula ordered a pizza with mushrooms and olives.

Unfortunately, the server's handwriting was quite sloppy. When the chef tried to read the order, he could see that Paula wanted two different toppings, but he couldn't tell which two she wanted. The chef decided to pick two toppings at random.

1. How many different two-topping pizzas are possible?

2. What is the probability that Paula will get the pizza she ordered? What is the probability that she'll get something different?

3. Paula actually likes all of the toppings except sausage and pepperoni. What is the probability that she will get a pizza she likes? What is the probability that she will get a pizza she doesn't like?

0 to 1, or Never to Always

For each probability given below, think up two situations that have that probability.

In one of those two situations, the probability should be based on a theoretical model. In the other situation, the probability should be based on observed results. Be imaginative!

1. Probability = 0

2. Probability = $\frac{2}{7}$

3. Probability = 75%

4. Probability = 1

5. Probability = 2.3

6. Probability = .01

Pictures of Probability

Pictures are important tools for understanding and communicating about mathematics. You have seen that probabilities can be represented with numbers as fractions, percentages, or decimals. You will now use pictures to represent probabilities.

Connecting pictures or diagrams to the probabilities they represent will help you work through the activities in this part of the unit. You will blend geometry with probability, creating area models to visualize the probabilities of the possible **outcomes** for an **event.**

You will use an area model to find the answers to questions about the results of rolling two dice. These pictures of probability will become a tool to help you evaluate strategies for the game of Pig.

Vanessa McDaniel and Randy Luiz play Linear Nim with colored cubes.

Rug Games

Each diagram in this activity represents a rug. Imagine that a trap door opens directly over the rug. A dart falls down through the door, landing at random somewhere on the rug. In other words, every point on the rug has the same chance of getting hit as every other point.

1. Suppose you were trying to predict the color the dart will hit. Which color would you choose, gray or white? What is the probability the dart will land on gray? What is the possibility it will land on white?

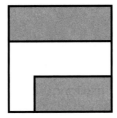

2. For each of rugs A through D, which color do you predict is most likely to be hit? For each color on each rug, find the probability of its being hit.

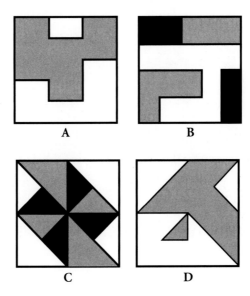

Portraits of Probabilities

Part I: Rugs for Events

Do these three tasks for each of the events numbered 1 through 5.

- State the probability of the event happening.
- Write a short explanation of why you believe you are correct.
- Draw a rug with a shaded portion that represents the probability.

1. Rolling a 6 with one die

2. Being chosen as one of three POW presenters out of a class of 30 students

3. Flipping a coin twice and getting two heads

4. The sun rising tomorrow

5. Having no homework for the rest of the semester

continued ▶

Part II: Events for Rugs

The shaded part of each rug represents the probability of an event occurring. For each rug, describe an event that you believe has the probability of occurrence represented by the shaded part of the rug.

6.
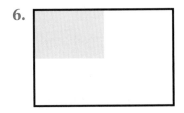

7.
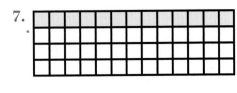

8.
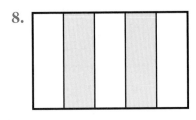

9.
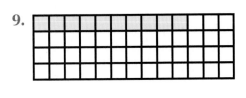

Linear Nim

Many strategy games involve two players who take turns removing objects from one of several piles according to certain rules. The winner is the person who removes the last object.

These games often go by the name Nim.

In one version of Nim, there is only one pile. In this case, the objects can be represented by a single row of marks on a sheet of paper. We will call this game Linear Nim.

Here's how a particular form of Linear Nim works.

At the beginning, there are 10 marks on a sheet of paper.

Each player, in turn, crosses out 1, 2, or 3 of the marks. It doesn't matter which marks are crossed out.

Play continues until all of the marks have been crossed out. The player crossing out the last mark is the winner.

○ Part I: Finding a Strategy

Your first task is to find a winning strategy for this particular game. You might begin by finding a partner and playing the game for a while.

As you and your partner play, pay attention not only to who wins, but also to when you realize who is going to win and how you know.

The question of who wins may depend on which player goes first. Therefore, one element of your strategy might be to decide whether you want to go first or second.

continued ▶

Be sure the strategy you develop is complete. That is, you should take into account every possible move that your opponent might make.

○ Part II: Variations

Once you have developed your strategy, investigate how it would have to change if the game were to vary.

This game starts with 10 marks and allows a player to cross out up to 3 marks at a time. What if these numbers were changed? For example, suppose you start with 15 marks or allow a player to cross out up to 4 marks at a time? You can change the *initial number* of marks, and you can change the *maximum per turn* that a player can cross out.

How would your strategy change if you varied the game? Are there some cases when you should choose to go first and others when you should choose to go second? What does it depend on?

Consider a variety of examples and look for generalizations.

○ Write-up

1. *Process:* Describe how you went about understanding the original game and developing a strategy. Indicate the key insights that were important in your understanding.

2. *Strategies*
   a. Describe the strategy you developed for the original game.
   b. Describe some specific variations you looked at and the strategy you developed for each.

3. *Generalizations:* State any general principles you developed about variations on Linear Nim. In particular, can you describe, in terms of the *initial number* and the *maximum per turn,* how to decide whether you'd want to go first or second?

4. *Self-assessment:* Explain what you learned. Be as specific as you can. Assign yourself a grade for your work on this POW. Explain why you think you deserve that grade.

Mystery Rugs

In each of the following situations, you are given some information about the probabilities of certain outcomes. For each situation, you will do two things.

• Make up a situation with outcomes that fit the given probabilities.

• Draw and label a rug that shows the probability for each outcome.

1. There are two possible outcomes for an experiment. One of the outcomes has a probability of $\frac{5}{12}$.

2. There are three possible outcomes for an experiment. One outcome has a probability of .4. A second outcome has a probability of .25.

3. There are three possible outcomes for an experiment. One outcome has a probability of 50%. The second outcome has a probability of 30%. The third outcome has a probability of 25%.

The Counters Game

Each player in this game needs a board with 11 boxes numbered from 2 to 12.

2	3	4	5	6	7	8	9	10	11	12

To start the game, each player puts 11 counters on a board. Players may place counters in the boxes in any way they choose, including putting more than one counter in a single box.

For each turn, a player rolls a pair of dice and then adds the numbers on the dice. Each player who has any counters in the square corresponding to that sum removes *one counter* from that square. Even if a player has more than one counter in that square, only one counter is removed. If a player has no counters in that square, the player does nothing. It doesn't matter who rolls the dice each time.

The winner of the game is the first player to remove all the counters from his or her board.

The challenge of the game is choosing where to place the counters so that you will likely be able to remove them quickly during the game.

1. Play one or two practice games with your group. Just guess where to place the counters.

2. Now think about where to place the counters. Write a sentence or two explaining what you think would be a good way to place them and why.

3. Play the game several times with your group, with each player using his or her own strategy for placing the counters.

4. To prepare for a competition among all the groups, discuss with your group the various strategies you used. Choose one strategy for the competition and record what it is.

Rollin', Rollin', Rollin'

Roll a pair of dice 50 times. With each roll, find the sum of the dice. Keep a record of your sums in an organized way.

1. Draw a **graph** of the data you gathered.

2. Write a paragraph about your results. You should summarize your observations about the data and discuss why the results came out the way they did.

3. What new thoughts does this experiment give you about how to play the counters game?

The Theory of Two-Dice Sums

You have played a game that involved two-dice sums. You have also done some experiments to get an idea of how two-dice sums are distributed.

Now it's time to look at the theory and to get more precise information about this distribution.

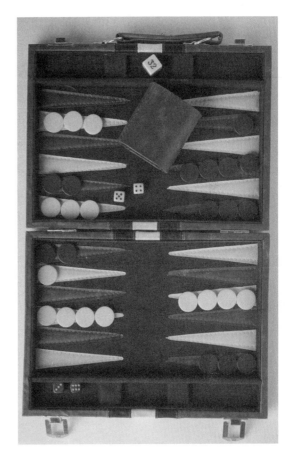

Work with your group to develop a rug diagram that will help you understand two-dice sums. The rug diagram will give you a theoretical model for finding the probability of each possible sum.

Keep in mind that equal **areas** of your rug should represent equally likely outcomes. You should assume that the dice are **fair.** That is, for a single die, the probability of getting each possible result is $\frac{1}{6}$.

You may find it useful to use two dice, preferably of different colors, as you work on your rug diagram. Think about the portion of your rug that corresponds to a given result for the pair of dice.

Money, Money, Money

1. Nina was working on the question from *What Are the Chances?* that asks for the probability of getting one head and one tail when a coin is flipped twice.

   She said that because there are exactly three possible outcomes— two heads, one head and one tail, and two tails—the probability of getting one head and one tail is $\frac{1}{3}$.

   Explain why she's wrong. Make your explanation as clear as you can, using diagrams as needed.

2. Imagine that you have two pockets in your jacket. Each pocket contains a $1 bill, a $5 bill, and a $10 bill.

   You reach in and remove one bill from each pocket. Assume that, for each pocket, the three bills are equally likely to be removed.

   a. What are the possible totals of the two bills you take out?

   b. What is the probability that the two bills will total exactly $2?

   c. What is the probability for each of the other possible totals?

3. What do Questions 1 and 2 have in common? How are these questions related to the problem of two-dice sums?

Two-Dice Sums and Products

Suppose you roll a pair of dice and add the numbers you roll.

1. Which is more likely: the sum is an odd number, or the sum is an even number? Explain why.

2. Make up two new probability questions about two-dice sums that can be answered from the rug diagram for two-dice sums. Answer your two questions if you can, and explain your answers.

Now suppose that instead of adding the numbers rolled on two dice, you *multiply* them. Let's call the result a *two-dice product.*

3. What are the possible two-dice products? What is the probability of getting each of the two-dice products?

4. Which is more likely: the product is an odd number, or the product is an even number?

5. Make up two probability questions about two-dice products. Answer them if you can.

In the Long Run

In recent activities, you have looked at ways of finding the probability of a given outcome, such as flipping two heads in a row or getting a sum of 9 on a pair of dice. You've flipped coins or rolled dice to determine the probability of a particular outcome. You've used rug diagrams to analyze probabilities. And you have confirmed some of your results with experimental evidence.

In the game of Pig, the probabilities are pretty simple, because each number on the die is equally likely to occur. But the situation is complicated by the fact that the payoff differs with each possible result, ranging from a gain of 6 points to a loss of all points.

How do you use probability to analyze what will happen in the long run in a game like this? In the next group of activities, you will first look at simple situations involving the long run. Gradually, you will explore more complex problems. You're moving toward your goal of finding the best strategy for Pig!

Eric Zafaralla, Santiago Saucedo, Maria Balderas, and Sergio Lopez play The Game of Pig.

Spinner Give and Take

Al and Betty are playing a game with this spinner.

Each time the spinner comes up in the white area, Betty wins $1 from Al. Each time the spinner comes up in the shaded area, Al wins $4 from Betty.

Al wins $4 from Betty.

Betty wins $1 from Al.

1. Who will come out ahead over the long run? Write down your prediction and explain your reasoning.

2. Now play the game for 25 spins. Write down your results.

3. If Al and Betty play 100 games, how far ahead is the expected winner likely to be?

You might make a spinner using a pencil and a paper clip. Bend open one end of the paper clip. Use the pencil to hold the other end in place as you spin the paper clip.

Pointed Rugs

In *Rug Games,* you decided which color in each of the rugs was most likely to be hit by a falling dart. Now you will work with rug diagrams again.

This time, you earn a certain number of points if the dart lands on the color you have chosen. This means that your choice of color involves more than just finding probabilities. You must also take into account the number of points that are awarded each time the dart lands on a certain color.

For each rug, decide which color is the best to bet on to maximize your points in the long run. Write clear explanations to support your answers.

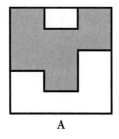

A

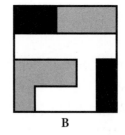

B

Gray: 6 points   Gray: 10 points

White: 8 points   White: 8 points

                             Black: 16 points

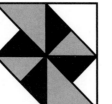

C

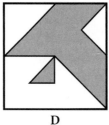

D

Gray: 5 points   Gray: 15 points

White: 6 points   White: 13 points

Black: 10 points

What's on Back?

This game involves three cards.

- One card has an **X** on both sides.
- One card has an **O** on both sides.
- One card has an **X** on one side and an **O** on the other side.

The three cards are placed in a bag. You draw one card at random and look at only one side. You cannot look at the other side or at the other cards. The goal of the game is to predict whether there is an **X** or an **O** on the other side of the card that you drew.

There are many strategies for making this prediction—some good, some not as good. Here are two possible strategies.

- Predict that the mark on the other side will be different from the mark you see. (That is, if you see **X**, predict **O**; if you see **O**, predict **X**.)
- Always predict that the mark will be an **X**.

No strategy will be successful all of the time. Try to find the probability of success for each strategy you consider.

Your ultimate goal is to find the strategy with the highest possible probability of being right.

For each strategy you consider, do these two tasks.

- Find an *experimental estimate* of the probability of success using that strategy. That is, devise an experimental method of testing your strategy. It will probably be useful to make a set of cards to do this.

    You will need to repeat your experiment quite a few times to get a good experimental estimate. You may want to repeat it until your overall results begin to stay roughly the same.

- Analyze the probability of success for that strategy by using a *theoretical model.*

continued ▶

Begin with the two strategies described earlier. Look for both an experimental estimate and a theoretical model for the probability of predicting correctly with each strategy.

Then think of other strategies. Try to determine the strategy with the highest probability of success for correctly predicting "what's on back."

○ *Write-up*

Your write-up should contain four parts.

1. *Problem Statement*

2. *Process:* Include a description of exactly what you did to carry out the experiments for part a.

3. *Results:* Describe each strategy you tried. For each, tell what you found as the probability of predicting correctly. If possible, describe your results in terms of both your experimental results and your theoretical analysis.

   State what strategy you think gives the highest probability of predicting successfully. Justify your answer.

4. *Self-assessment*

Mia's Cards

1. Mia is playing a game that involves picking a card from a standard deck. A standard deck has 52 cards, with 13 cards in each of 4 suits. The suits are clubs, diamonds, hearts, and spades. The 13 cards in each suit are the ace, 2, 3, 4, 5, 6, 7, 8, 9, 10, jack, queen, and king.

   Mia mixes up the cards and then picks a card at random from the deck. She gets 10 points if the card is a heart and 5 points if it's a club, spade, or diamond. Then she puts the card back in the deck.

   Suppose she does this many times. What will be her average number of points each time she chooses a card? Explain your answer.

2. On February 14, Mia changes the game so that she gets 20 points for a heart, 15 points for a diamond, and no points for a club or a spade. If she plays this new game many times, what would you expect her average score per card to be?

3. Now create and analyze your own game.

   a. Make up a game like Mia's, in which a person picks a card and receives a number of points that depends on the type of card picked.

   b. Calculate the average score per pick in the long run for your game.

A Fair Rug Game?

1. Tony and Crystal are sitting around this rug watching darts randomly fall from the ceiling.

   If a dart lands on the white part of the rug, Crystal wins $5 from Tony. If a dart lands on the black part, Tony wins $3 from Crystal.

   Do you think this is a fair game? What is Tony's **expected value** for each turn? What is Crystal's? That is, what is each player's average score per turn in the long run?

2. If you think the game is not fair, make it fair by changing the amount of money each player wins. Don't change the rug.

One-and-One

Sometimes in a basketball game, a player is presented with a situation known as a *one-and-one.*

In a one-and-one situation, the player begins by taking a free throw at the basket. If the player misses, that's the end. But if the shot is successful, the player gets to take a second shot.

One point is scored for each successful shot. So the player can end up with 0 points (by missing the first shot), 1 point (by making the first shot but missing the second), or 2 points (by making both shots).

Over many games, Terry has shown that whenever she attempts a free throw, she has about a 60% probability of making it.

> In a one-and-one situation, how many points is Terry *most likely* to score: 0, 1, or 2?

Write down your intuitive guess to this question. Explain why you think that might be the answer.

A Sixty Percent Solution

Terry is in a one-and-one free throw situation. As in *One-and-One,* she has a 60% probability of making any given shot.

1. Devise a way to simulate Terry's one-and-one probabilities. Describe your **simulation.**

2. Do your simulation of Terry's one-and-one situation 40 times. State your results.

3. What was the *most frequent outcome* in your simulation?

4. What was the *average score* per one-and-one situation?

The Theory of One-and-One

You've made guesses about it and done simulations of it. Now it's time to work out the theory.

Once again, Terry has a 60% chance of making a free throw. She is in a one-and-one situation.

Develop a theoretical analysis. Create a rug diagram to picture the probabilities for each score: 0, 1, or 2. What is her expected score in a one-and-one situation?

A rug diagram is more formally called an **area model.**

Streak-Shooting Shelly

When Streak-Shooting Shelly steps up for a one-and-one situation, her chances of making the first shot are 80%. If she makes her first free throw, there is a 90% chance she will make her second free throw.

1. In what percentage of one-and-one situations will Shelly score no points? One point? Two points?

2. What is Shelly's expected value per one-and-one situation?

Spins and Draws

1. Al and Betty are playing spinner games again. This time the spinner is divided so that the arrow will land in Al's area $\frac{1}{5}$ of the time and in Betty's area $\frac{4}{5}$ of the time.

   Al pays Betty 30¢ when the arrow lands in her area. Betty pays Al $1.25 when it lands in his area.

   a. What is Al's expected value per spin? What is Betty's?

   b. How might the payments be changed so that the game is fair?

2. Ari and Brenna are playing a game that involves drawing a card from a standard deck. After each draw, the card is returned to the deck. It doesn't matter which person draws the card—all that matters is which card is drawn.

   If the card drawn from the deck is a jack, Brenna pays Ari 20¢. If the card drawn is a heart, Ari pays Brenna 8¢. If neither a jack nor a heart is drawn, Ari and Brenna each give a penny to charity.

   What is the expected value per draw for Ari? For Brenna? For the charity?

Aunt Zena at the Fair

Aunt Zena has gone to the weekend fair put on by her nephew's school. The school is trying to raise funds so it can offer some special classes.

Aunt Zena's favorite game is the ring toss. The goal is to toss a large ring onto a stick.

Each time Aunt Zena succeeds, she wins a coupon for a free dinner, donated by a local restaurant. She figures the coupon is worth about $12. It costs Aunt Zena $1 for each toss.

1. Aunt Zena spent most of Saturday afternoon at the ring toss booth. She was able to win the ring toss only about once every 20 tries.

   If she continued like this, would she win or lose money in the long run? Consider each coupon she wins as the equivalent of $12.

   What would be her expected value per toss?

2. Aunt Zena went home determined to do better the next day. She practiced late into the night. On Sunday, she returned to the fair and won the toss about once every 10 tries.

   If she continued like this, would she win or lose money in the long run? What would be her expected value per toss?

Simulating Zena

Your graphing calculator can generate random numbers. For example, it can randomly generate a decimal number between 0 and 1. The calculator will show a fixed number of decimal places, often 10. For example, it might give these numbers: 0.9435974025, 0.1466878292, 0.4058096418.

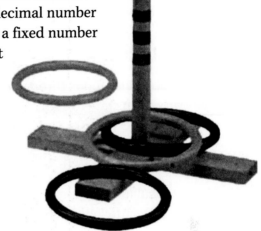

In the long run, for numbers chosen at random, 10% of the numbers will be between 0 and 0.1, 10% will be between 0.1 and 0.2, and so on. Half the numbers will be between 0 and 0.5, and two-thirds of the numbers will be between 0 and 0.6666666667.

Use your graphing calculator to create a simulation for Question 1 of *Aunt Zena at the Fair*.

> Aunt Zena is playing ring toss. It costs her $1 to play. If she wins, she gets $12. Her skill allows her to win about once every 20 tries. In the long run, how much will she win per try?

1. Describe your simulation.

2. Use your simulation to test your theoretical analysis of her expected value. How do they compare?

The Lottery and Insurance—Why Play?

In this activity, you will look at two real-life situations that involve probabilities to see if expected value tells the whole story.

1. The Lottery

Many states raise funds through lottery games. The rules vary from state to state. For this activity, assume each lottery ticket costs $1. Suppose that for a certain week, about 14 million tickets were sold, and the winning ticket was worth $6 million.

a. Calculate the approximate expected value of a lottery ticket that week.

b. Do you think buying a lottery ticket is a wise investment? Explain your answer.

Many state lotteries use a portion of the proceeds to fund education.

2. Insurance

Buying insurance can be thought of as similar to playing a lottery. You pay a certain amount, called the *premium*, to the insurance company. Most of the time, the insurance company just gets to keep your money and pays you nothing.

Sometimes, however, you have a claim that is covered by insurance. When that happens, the insurance company has to pay for your losses. In general, they pay you much more than you paid as a premium. You "win" whenever you collect on your insurance. Of course, you don't want to be in an accident or fire. The insurance game is a game you don't really want to win.

In the long run, insurance companies take in more money in premiums than they pay out in claims, or they wouldn't be in business. In other words, the expected value for the insurance company is positive, and yours is negative.

So why do people play?

Martian Basketball

In Martian basketball, instead of having one-and-one free throw situations, they have one-and-one-and-one situations.

In other words, if a player makes both the first and second shots, he or she can take a third shot. So the player can get 0 points, 1 point, 2 points, or 3 points.

1. Suppose our friend Terry (from the original one-and-one problem) has moved to Mars and is playing basketball there. At home, her probability of success was the same for every free throw. Assume that this is still true on Mars.

   However, suppose her probability of success on each shot is down to 40% because she is adjusting to the gravity on Mars.

   a. What is Terry's probability of getting each possible score in a one-and-one-and-one situation?

   b. How many points is she most likely to score in a one-and-one-and-one situation?

   c. What is her expected value for each one-and-one-and-one situation?

2. Now suppose Streak-Shooting Shelly has moved to Mars. Like Terry, Shelly is still adjusting to Mars. Her overall performance is down, but she still shoots better when she has just made a shot.

   Suppose she has a 60% probability of making her first shot. If she gets the first shot, she has an 80% probability of making the second shot. If she gets the first two shots, her probability of making the third shot rises to 90%.

   a. How many points is Shelly most likely to score in a one-and-one-and-one situation?

   b. What is her expected value for each one-and-one-and-one situation?

The Carrier's Payment Plan Quandary

During the last century, newspapers were often delivered by a carrier who had already paid for the papers. The carrier then collected payment from the customer and kept whatever he or she collected.

Suppose one day a customer said to the carrier, "Instead of collecting the usual $5 per week, how about if you just pick two bills at random out of this bag? You get to keep whatever you pick. If you choose to pick out of the bag, you'll do that every week from now on."

The customer showed the newspaper carrier the bag, which contained one $10 bill and five $1 bills. Thus, two sums were possible: $11 or $2. The customer put in new bills each week to replace the ones taken the week before.

Should the carrier accept the customer's offer?

1. Plan and carry out a simulation for a reasonable number of trials. Based on your simulation, decide which is the better choice for the carrier. Explain your decision.

2. Now use an area model or other method to compute the carrier's expected value from the alternative payment plan. That is, find the average the carrier would expect to get per week in the long run using this payment plan. Based on your theoretical analysis, decide which is the better choice. Explain your decision.

3. Which method do you trust more, the simulation or the theoretical analysis? Why?

A Fair Deal for the Carrier?

This problem is a simpler version of the situation in *The Carrier's Payment Plan Quandary*. As in the original situation, the customer would ordinarily pay $5 per week for newspapers.

In this problem, the customer places a $20 bill and four $1 bills in a bag. The carrier's option is to draw out just one bill at random from the bag.

1. Imagine that you are the paper carrier. Figure out whether this option would give you a better-than-fair deal in the long run. In other words, is the expected value for this option better than the usual $5 per week? Explain how you arrived at your decision.

2. Devise some way to simulate this alternate payment plan without the use of technology. Carry out 20 trials. Find the average result of your trials.

3. Think about how you might create a technology-based simulation of the alternate payment plan that will allow you to do many trials quickly. Write down your ideas.

Simulating the Carrier

Now you will create and use a technology-based simulation for the simplified problem in *A Fair Deal for the Carrier?*

Part 1: Creating the Simulation

Here is a summary of the situation.

> A customer puts a $20 bill and four $1 bills in a bag. The carrier pulls one bill at random from the bag as payment for newspaper delivery.

Use a calculator or a computer to create a simulation for this situation. Your simulation should be able to execute the process many times and calculate the carrier's average result.

Part 2: Using the Simulation

Now examine what you can learn from your simulation.

1. Run the simulation several times, using the same number of trials. How do your results vary from one run to the next? How do they compare with the theoretical prediction?

2. Now do several runs with a larger number of trials. Compare your answers with each other and with the theoretical prediction. How are these results different from those in Question 1?

3. What conclusions can you draw about the influence of the number of trials on the results?

4. Does your simulation give you more confidence in the theoretical analysis? Explain.

Another Carrier Dilemma

Here is one more variation on the newspaper carrier's situation. This time the customer places two $5 bills and three $1 bills in a bag and allows you (the paper carrier) to draw out two bills at random.

If you use this method week after week, how much would you expect to earn each week, on average, in the long run?

Explain your reasoning. Use more than one method to analyze the situation.

Analyzing a Game of Chance

In *One-and-One,* Terry took at most two shots at the basket in each one-and-one situation. In *Martian Basketball,* each turn involved at most three shots.

In the game of Pig, there is no limit to how long a turn can last. A person could roll 20, 50, or even 100 times without getting a 1 on the die, continuing to earn more points. This is one reason the game is so complicated to analyze.

Often a good approach to solving a complicated problem is to examine a simpler, but related, problem. That's what you will do next, using a simplified version of Pig, called Little Pig.

You'll play for a while to get used to this new game. Then you'll start analyzing strategies by using area models.

When you are finished with Little Pig, you'll apply your new insights to find the best strategy for the original game of Pig.

Then you'll be ready to wrap up the unit with portfolios and end-of-unit assessments.

Edward Rokos prepares his analysis of Little Pig.

The Game of Little Pig

To play Little Pig, you need a bag containing three cubes—one red, one blue, and one yellow. Instead of rolling a die as you did in Pig, you will draw a cube out of the bag.

In each turn, you can draw as many times as you want, replacing the cube after each draw, until you decide to stop or you draw a yellow cube. Each time you draw a red cube, you add 1 point to your score. Each time you draw a blue cube, you add 4 points to your score.

If you stop before you draw a yellow cube, your score for that turn is the total number of points for all draws in that turn. But if you draw a yellow cube, the turn is over, and your score for that turn is zero.

Your eventual goal will be to find a strategy for Little Pig that will give the highest possible average score per turn in the long run. In other words, you want the strategy with the highest possible expected value per turn

For now, you will informally investigate the game.

1. Play Little Pig several times in your group, noting different possible strategies.

2. Make a list of some possible strategies for playing Little Pig.

3. As a group, choose a single strategy that you think might give the best results. Be sure to write this strategy clearly.

Pig Tails

The game of Pig Tails is another variation on the game of Pig. Here's how you play.

Each turn consists of flipping a coin until either you decide to stop or you get tails. If you stop before getting tails, your score for the turn is the number of heads you got—that is, the number of times you flipped.

But if you get tails before deciding to stop, your score for the turn is zero.

1. What is your expected value per turn if your strategy is to flip just once and then stop, no matter what the result?

2. Now consider the strategy of always flipping twice (unless you get tails on the first flip) and then stopping. What is your expected value for this strategy?

3. What is the expected value per turn if you always flip three times (unless you get tails on the first or second flip)?

4. Generalize Questions 1 through 3. That is, find the expected value per turn if your strategy is to flip n times and then stop (unless you get tails on an earlier flip).

Little Pig Strategies

Now that you have some experience playing Little Pig, it's time to analyze some strategies.

Two strategies are described below. For each strategy, use an area model to describe what might happen and to find the expected value per turn using that strategy. Remember these rules.

- Draw red to earn 1 point.
- Draw blue to earn 4 points.
- Draw yellow and lose all points in the current turn.

The 2-Point Strategy

In this strategy, you stop as soon as you have at least 2 points. Of course, if you draw a yellow cube before getting 2 points, you'll have to stop sooner.

The 2-Draw Strategy

In this strategy, you stop after drawing two cubes, no matter what the results of those two draws. Of course, if the first cube is yellow, you'll have to stop after only one draw.

Continued Little Pig Investigation

You have already found the expected value for some Little Pig strategies.

Continue your investigation. You will use strategies of your own choice, or your teacher will assign you specific strategies to investigate.

As you work, keep in mind the ultimate goal of finding the strategy with the greatest possible expected value per turn.

Should I Go On?

1. Suppose there are two classes, with 30 students in each class.

   In both classes, the students are individually playing Little Pig. In both classes, every student has gotten exactly 10 points so far in the current turn.

   In one class, the students each decide to draw just one more cube. In the other class, they all decide to stop at 10 points.

   a. For the class in which students each draw once more, how many students would you expect to end up with 0 points? With 11 points? With 14 points? What would you expect for the total number of points in the class? What would you expect for the class average?

   b. Compare the expected class average from the class where students draw again with the average for the class in which everyone stops at 10 points. Which class would you expect to have a better average? Why?

2. Now suppose the situation is the same except that every student has only 2 points. Again, the students in the first class each draw just once more, while the students in the second class all stop at 2 points.

   a. What would you expect for the average score in the class where all the students draw once more?

   b. Which class would you expect to have a better average? Why?

The Best Little Pig

You've seen that in the long run, a person with 2 points in Little Pig will do better by drawing again, while a person with 10 points should stop.

So what is the best strategy for Little Pig? For what scores does it pay to draw again, and for what scores should you stop?

Based on your findings, what strategy will give the highest possible expected value per turn?

Big Pig Meets Little Pig

In *Should I Go On?* you looked at the question of when a player should draw again in Little Pig. Now you will examine how to apply that approach to Big Pig—that is, to the original game of Pig.

Assume there are two large classes of students. This time all the students are playing Big Pig.

1. Suppose each student has a current score of 10 points. In one class, each student rolls one more time and then stops. In the other class, each student stops at 10 points.

   Which class would you expect to end up with the better average score?

2. Consider at least two other initial scores (instead of 10 points). Decide which class you would expect to be better off—the class in which each student rolls once more, or the class in which all the students stop.

The Pig and I

You've worked through Little Pig strategies and then gone back to study the original game of Pig.

Based on your experiences with both games, summarize what you have learned about the best possible strategy for Pig.

Be sure to include what you think is the best strategy, how you found it, how you can **justify** that choice, and what you learned about "roll" strategies and "point" strategies.

Beginning Portfolio Selection

This unit involved two main approaches for finding probabilities.

- Simulations and experiments
- Theoretical analyses, using area models, tree diagrams, or other methods

Select an activity from the unit that represents each approach. Explain what each activity was about. Describe what you learned about probability from it.

This selection and explanation is the first step toward compiling your portfolio for this unit.

Natali Camacho and Irvin Vasquez work on their portfolios.

The Game of Pig Portfolio

It is time to put together your portfolio for *The Game of Pig*.

- Write a cover letter that summarizes the unit.
- Choose papers to include from your work in the unit.
- Discuss your personal growth during the unit.

Cover Letter

Look back over *The Game of Pig*. Describe the central problem of the unit and the main mathematical ideas. Your description should give an overview of how the key ideas were developed and how they were used to solve the central problem.

Select some activities that you think were important in developing the unit's key ideas. Your cover letter should include an explanation of why you selected each item.

continued ▶

Selecting Papers

Your portfolio for *The Game of Pig* should contain these things.

- *The Pig and I* and *Beginning Portfolio Selection*

  Include the two activities from the unit that you selected in *Beginning Portfolio Selection,* along with your written work on these activities.

- Other key activities

  Include two or three other activities that you think were important in developing this unit's key ideas.

- A Problem of the Week

  Select one of the three POWs you completed during this unit— *A Sticky Gum Problem, Linear Nim,* or *What's on Back?*

- Other quality work

  Select one or two other pieces of work that demonstrate your best efforts. These can be from any part of the unit, such as Problems of the Week, group or individual activities, and presentations.

Personal Growth

Your cover letter describes how the mathematical ideas develop in the unit. As part of your portfolio, write about your own personal development during this unit. You may want to address this question.

How might the ideas in this unit affect your own behavior in situations that involve probability?

Include any other thoughts you wish to share with a reader of your portfolio.

SUPPLEMENTAL ACTIVITIES

Probability, expected value, and the use of strategies are three important themes in *The Game of Pig*. The supplemental problems for this unit continue these themes. These are some examples.

- *Different Dice, Three-Dice Sums,* and *Heads or Tails?* ask you to find the probabilities for some events involving dice and coins.

- *Expected Conjectures* and *Squaring the Die* offer you the chance to gain more insight into the concept of expected value.

- *Counters Revealed* and *Piling Up Apples* concern strategies for the counters game you played and for a new game involving picking apples from piles.

Average Problems

1. Lucinda bought a dozen eggs on three different occasions. The average cost per dozen was $2.18.

   Give several possible combinations for what each of the three purchases might have cost.

2. Two classes took an exam. In the first class, the average score was exactly 78%. In the second class, the average score was exactly 86%. But when the two classes were treated as one large group, the average was not 82%.

   a. How is this possible?

   b. What's the highest that the combined average could be? What's the lowest?

   c. Under what circumstances would the average of the two class averages be the same as the average you get when you treat the two classes as one large group? Explain your answer as completely as you can.

   Be sure to justify your answers.

3. Garrison Keillor describes Lake Wobegon as a place where "the children are all above average."

   Suppose someone measured all 85 ten-year-olds in the town and found that their average height was 4 feet 7 inches.

   a. Is it possible that all the ten-year-olds are taller than 4 feet 7 inches?

   b. If not, what is the greatest number of ten-year-olds in the town who could be taller than that? Explain your answer.

   c. Is it possible that Garrison Keillor is right and all the ten-year-olds in Lake Wobegon are above average? What could that mean?

Above and Below the Middle

Suppose you combined the results from an entire class for *Waiting for a Double*. If you made a frequency bar graph of that information, it might look like the graph below. This graph represents a total of 300 trials (10 trials for each of 30 students). The numbers above each bar show how often a particular result occurred.

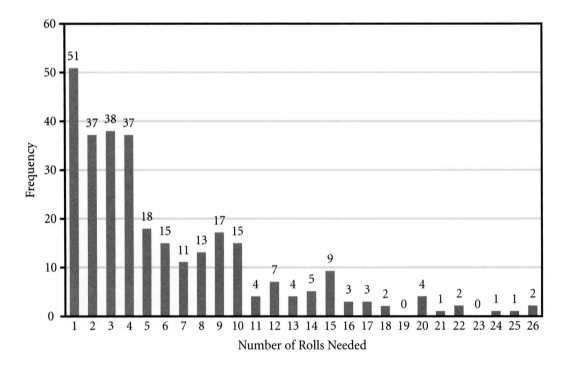

The first bar shows that in 51 trials, a double came up on the first roll of a pair of dice. The second bar shows 37 trials required two rolls of the dice to get a double; 38 trials required three rolls; and so on.

1. Find the **mean** (average) for the number of rolls needed, based on this graph.

2. How many of the results are above the mean? How many are below? Why are there so many more results below average than above average?

continued ▶

The mean is one type of middle value for a set of data. Another middle value is the **median.** You find the median of a set of data by listing the numbers in increasing order and finding the number halfway through the list.

The set of numbers 6, 2, 7, 4, 11, 8, and 9 is first arranged in increasing order:

2, 4, 6, 7, 8, 9, 11

The median is the number halfway through the list, which is 7. Three numbers are below 7 and three numbers are above 7.

If the number of terms in a list is even, split the difference between the two middle numbers. For instance, for the numbers 9, 3, 6, 2, 11, and 14, arrange them in increasing order:

2, 3, 6, 9, 11, 14

In this case, there are two middle numbers, 6 and 9. (There are two numbers below 6 and two above 9.) The median is the value halfway between 6 and 9, which is 7.5.

If a number appears more than once in a list, keep that repetition and arrange the numbers in increasing order. For example, arrange the list 2, 8, 3, 2, 3, 6, and 7 in increasing order:

2, 2, 3, 3, 6, 7, 8

The "middle" number in this list is the "second" 3 (there are three items to its right and three to its left); so the median is 3.

3. Find the median for the *Waiting for a Double* data in the frequency bar graph. Explain how you found it. You don't have to write out all 300 numbers to do this.

4. Which number—the mean, the median, or perhaps something else—do you think best represents what happens when you "wait for a double"? Explain your thinking.

Mix and Match

Glenn likes to wear gloves. But he has some funny habits about his gloves. He keeps all of his left-hand gloves in one drawer and all of his right-hand gloves in another.

When he gets ready to go out on a cold day, he just pulls out one glove at random from each drawer and puts them on, without checking to see if they match.

Today, his left-hand glove drawer has two brown gloves and three gray gloves. His right-hand glove drawer has one brown glove, two gray gloves, and two red gloves. (Glenn also loses a lot of gloves.)

1. What is the probability that he will pull out a pair of gray gloves today?

2. What is the probability that he will pull out a matching pair of gloves today?

3. What is the probability that neither of his gloves will be brown?

Flipping Tables

Suppose you flip a coin two or more times. Here are two ways to report the result.

- *As a combination:* For instance, if you flip three times, you might say, "I got heads twice and tails once."

- *As a sequence:* For instance, if you flip three times, you might say, "I got heads, then heads again, and then tails."

The difference is that with a combination, you don't report the order in which the results occurred. You simply say how many times you got heads and how many times you got tails.

For two flips, the lists below show all the possible combinations and sequences, using H for heads and T for tails.

Combinations	Sequences
2H	HH
1H and 1T	HT
2T	TH
	TT

The lists show that for two flips, there are three combinations and four sequences.

If you want to know the probability of a particular outcome, such as getting heads on both flips, you need to work with *equally likely* outcomes. As you have seen, the sequences are equally likely, so the chance of getting heads both times is one out of four, or $\frac{1}{4}$.

Now you will explore what happens for more than two flips.

1. Make an In-Out table in which the *In* is the number of flips and the *Out* is the number of different *combinations*. For just one flip, there are two combinations: 1H and 1T. The previous list shows that for two flips, there are three combinations. So start your table like this:

Number of flips	Number of combinations
2	3

continued ▶

a. Extend your table through at least six flips.

b. Find a rule to describe your table.

c. How many combinations are there for 10 flips? For 50 flips? For 100 flips?

d. Why do you think your rule is correct? Your explanation should explain the situation itself and not just show that the rule fits the results in your table.

2. Make another In-Out table. Again, the *In* is the number of flips, but this time the *Out* is the number of different *sequences*. For a single flip, there are two sequences, H and T. The list earlier in this activity shows that for two flips, there are four sequences. So the table for sequences starts like this.

Number of flips	Number of sequences
1	2
2	4

a. List all the sequences for the cases of three flips and four flips. This will require careful, organized work. Put your results into the table.

b. Find a rule or pattern to describe your table.

c. Based on your answer to part b, how many sequences would you expect for five flips? For six flips? For ten flips?

d. Why do you think your rule or pattern is correct? Remember to describe the situation and not just show that the rule fits the results in your table.

3. If you flip a coin four times, what is the probability that you will get heads every time? What if you flip five times? Six times? Ten times?

4. If you flip a coin four times, what is the probability that you will get three heads and one tail? You might use your work from earlier in this activity. How many of the *equally likely* outcomes give this result?

A Pizza Formula

The pizza place in *Paula's Pizza* offered six toppings: sausage, onions, mushrooms, pepperoni, olives, and peppers. You found the number of different two-topping pizzas that can be made from these six toppings.

What if the pizza place offered fewer toppings or more toppings? How would the number of different two-topping pizzas change?

1. Make a table in which the input is the number of toppings available and the output is the number of different two-topping pizzas that can be made.

2. Describe as precisely as you can any patterns you see in your table. If possible, explain why the patterns you find must continue throughout the table.

3. Look for a formula, or rule, for computing the output from just the input. For example, your rule should allow you to figure out how many two-topping pizzas could be made if 500 different toppings were available.

4. Find an explanation for your rule in terms of the situation.

Heads or Tails?

You know that if you toss a fair coin many times, you will get heads about half the time and tails about half the time.

Suppose you toss two fair coins. The diagram shows that the probability of both coins coming up heads is $\frac{1}{4}$.

Coin 1

	H	T
H	Both heads	Tails on coin 1; heads on coin 2
T	Heads on coin 1; tails on coin 2	Both tails

Coin 2

For example, if 100 people each tossed a pair of coins, about 25 of them would get two heads. Similarly, about 25 would get two tails, and about 50 would get one head and one tail.

1. Do a similar analysis for the case of three coins. That is, suppose three coins are tossed at once. Find each of these probabilities.

    a. The probability of getting heads on all three coins

    b. The probability of getting two heads and one tail

    c. The probability of getting one head and two tails

    d. The probability of getting tails on all three coins

2. Now explore the case of four coins. You might begin by making a list of the possible results.

3. Suppose you flip ten coins.

    a. What is the probability of getting all heads?

    b. What is the probability of getting nine heads and one tail?

4. What generalizations can you make about your results?

Small Nim

The version of Linear Nim described in the POW *Linear Nim* involves an initial setup with ten marks. Now you will explore some simpler cases of the game.

1. Consider the version of the game that begins with only four marks.

   As before, each player, in turn, crosses out one, two, or three of the marks.

   Imagine that you are the second player. Plan how you will respond to each possible move by the first player.

   a. What will you do if the first player crosses out just one mark?

   b. What will you do if the first player crosses out two marks?

   c. What will you do if the first player crosses out three marks?

2. Now consider the case in which there are five marks at the beginning of the game.

   Plan how you will respond to each possible move by the first player.

3. If you are the first player, and the game begins with five marks, can you be sure to win? If so, how will you do it? If not, why not?

Counters Revealed

Here is a simplified version of the counters game.

As in the regular version, the game board consists of a rectangle with squares numbered from 2 through 12.

2	3	4	5	6	7	8	9	10	11	12

But instead of 11 counters to place on the board, you have only 2. You can choose to place them in the same square or in two different squares.

As before, a pair of dice is rolled on each turn, and the numbers are added. If you have a counter in the square that matches the sum, you remove it (but only one counter per roll). Your goal is to remove your counters as quickly as possible.

The challenge is to decide where to place the two counters to make your chances of winning as good as possible.

1. Decide on at least three ways of placing the counters that you'd like to test.

2. Set up a game board for one of your choices.

3. Play the game. See how many rolls it takes to win. Keep repeating until you think you have a good idea of how many rolls it would take, on average, to win with that placement of the counters.

4. Now repeat Steps 2 and 3 for each of your other choices of initial position.

5. Write a report describing your work and summarizing your results. State any conclusions you reached, and explain them as well as you can.

Two-Spinner Sums

In some board games, you move your marker a certain number of spaces based on the result of a spinner. Suppose you were playing a game using these two spinners.

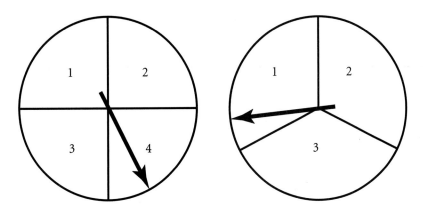

The rule is that you spin each spinner and move the same number of spaces as the sum of the two results.

1. Suppose you need a total of at least 5 from the two spinners to reach the winning position on the board. What is your probability of success?

2. Consider the various possible two-spin sums and the probabilities for each. How do these results compare with the results for two-dice sums? What are the similarities? What are the differences?

Different Dice

Imagine a set of dice in which every 4 was replaced by a 7. Each die could roll 1, 2, 3, 5, 6, or 7, with each result equally likely.

Find the probability of each of the following results when rolling two of these dice.

1. The sum of the dice is 7.

2. The sum of the dice is less than 7.

3. The sum of the dice is greater than 7.

4. The product of the dice is even.

5. The product of the dice is odd.

6. The numbers on the dice are the same.

7. The sum of the dice is a multiple of 3.

8. The product of the dice is a multiple of 3.

Three-Dice Sums

Suppose you roll three standard dice and then add the results. The lowest sum you can get is 3 (by rolling three 1's), and the highest is 18 (by rolling three 6's).

1. Without doing any analysis, what sums would you expect to be the most likely? Why?

2. Find the probability of getting each of the possible three-dice sums. Describe any patterns you find, and explain them if you can.

3. Make up and answer some questions about three-dice sums.

Piling Up Apples

One afternoon, Al and Betty were collecting apples. They had gathered them into two piles and decided to play a game with them.

They would take turns taking some apples from one of the piles. The one who took the last apple would win. They decided on these rules.

- On any turn, a player must take at least one apple.

- On any turn, a player can take apples from only one of the piles.

At first, they decided to go alphabetically, so that Al would always choose first. At the end of each game, they would put the piles back the way they originally were.

They began with 11 apples in Pile A and 8 apples in Pile B. After a few games, Al found a strategy so that he always won.

1. Figure out Al's strategy, and explain how he could always win.

The next day they played the same game, except that the piles of apples were not the same sizes as the day before. Al still insisted on going first, but now, no matter what he did, Betty always won.

2. Figure out how many apples might have been in each pile so that Betty could always beat Al, and explain how she won.

Each morning after that, the friends decided what size piles they would use for that day. On each day, either Al won every game or Betty won every game. Al still always went first.

continued ▶

3. Represent this situation with a rule.
   a. Find a rule, in terms of the sizes of the piles, that explains which player will always win.
   b. For the sizes where Al always wins, explain his strategy.
   c. For the sizes where Betty always wins, explain her strategy.

One day they decided to put the apples into three piles. Al still always went first, and they kept the same rules.

They discovered that with three piles, it was harder to predict who would win.

4. Find sizes for three piles so that Al can always win. Explain how he should play to ensure a win.

5. Find sizes for three piles so that Betty can always win. Explain how she should play to ensure a win.

Make a Game

Work with a partner to invent a game that uses probability and strategy.

Write clear instructions so that other people can understand how to play your game. Test your instructions several times with someone at home to make sure they are easy to follow.

Your written work should include instructions for your game, strategies for winning, and an explanation of how you used ideas about probability and strategy in designing and analyzing your game.

Pointed Rug Expectations

In *Pointed Rugs,* you looked at several rug diagrams. Each rug had two or more colors. For each color, you earned a certain number of points if a random dart landed on that color. Your task was to determine which color you should bet on to maximize your points in the long run.

In this activity, you will use the same rugs, with the same point scores, but the task is different. For each rug, answer this question.

Suppose you drop a dart many times on the rug. Each time, the dart lands in a random spot on the rug. What is the average number of points you will get per dart? In other words, what is the expected value for this rug?

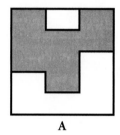

A

B

Gray: 6 points   Gray: 10 points

White: 8 points   White: 8 points

                 Black: 16 points

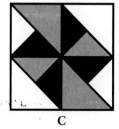

C

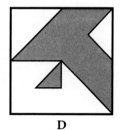

D

Gray: 5 points   Gray: 15 points

White: 6 points   White: 13 points

Black: 10 points

Explaining Doubles

In *Waiting for a Double,* you rolled a pair of dice repeatedly to see how many rolls it took to get doubles. You rolled for doubles many times and examined the results. One part of that experiment concerned the *average,* or *expected value,* of the number of rolls to get a double.

Explain, based on theoretical probability, the expected value for waiting for doubles. You might pick a large number and imagine rolling your pair of dice that many times. How many doubles would you expect to get? What does that say about the average number of rolls needed to get a double?

Two Strange Dice

These dice may look ordinary, but they aren't. Each die has three sides not shown here. Those six "invisible" sides have just one dot each. In other words, each die has four 1s, a 4, and a 5.

In a certain game, a player is offered a choice.

- Get 5 points.
- Roll this pair of dice, add the results, and get that number of points.

If you were to play this game many times, which choice would give you more points in the long run? Explain your answer thoroughly.

Expected Conjectures

Al and Betty were getting used to the idea of
expected value, and they were making
some conjectures.

Al wanted to find the expected
value from rolling one die. He
imagined rolling 600 times and
figured that he would get about
100 one's, 100 two's, and so on. So
he first did this computation:

$$100 \cdot 1 + 100 \cdot 2 + \cdots + 100 \cdot 6$$

This gave a total of 2100 points for
the 600 rolls. He then divided by 600
to get the average per roll, which came out to 3.5.

1. Betty tried it with 6000 rolls and got the same average. Explain why
   their averages are the same. (Look for more than one explanation.)

2. When Al saw that the average was the same both ways, he decided
   he could find the average with only 6 rolls. Would he still get the
   same result? Explain.

3. Now use the variable *N* for the number of rolls. Use algebra to prove
   that you get the same average per roll no matter what the value
   of *N* is.

4. Could you find the average with only one roll? Explain.

Squaring the Die

Here are some more conjectures that Al and Betty are making about expected value. What do you think of their ideas?

Use the definition of *expected value* based on "the long run" to justify your evaluation of their conjectures.

1. Al says, "If you roll a die, the expected value is 3.5, because results of 1 through 6 are equally likely, and 3.5 is the average of the numbers 1, 2, 3, 4, 5, and 6."

   Does this give the right answer? Explain.

2. Betty says, "If you roll two dice, the expected value should be twice as big as if you roll one die." Based on Al's idea in Question 1, she thinks the expected value for a two-dice sum is 7.

   What do you think? Explain.

3. Al says, "If the expected value for the *sum* of two dice is 3.5 + 3.5, then the expected value for the *product* of two dice should be 3.5 · 3.5, which is 12.25."

   Is Al right? Explain.

4. Finally, Betty says, "If you roll a die and square the number that shows, the expected value for the result should be the square of the expected value for a single die—that is, 3.5^2, which is 12.25."

   Is Betty right? Explain.

Fair Spinners

1. Al and Betty are interested in changing their spinner game so that neither player has an advantage.

   They decide to use the spinner from *Spinner Give and Take,* but they want to modify the game by changing the amount that Al wins when the spinner lands on the shaded area.

   In other words, Betty still wins $1 from Al when the spinner lands in the white section. But Al wins some other amount when the spinner lands in the shaded section.

   What should the new amount be so that the game is fair? Explain how you arrived at your answer.

2. Make up and solve a spinner problem of your own.

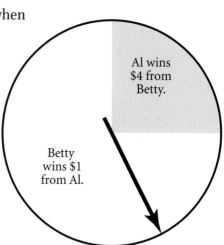

Al wins $4 from Betty.

Betty wins $1 from Al.

Free Throw Sammy

Sammy has a free throw success rate of 80%.

Construct an area model for the one-and-one situation with Sammy.

1. Using your area model, what is the probability that Sammy will get
   a. 0 points?
   b. 1 point?
   c. 2 points?

2. Use your area model to find Sammy's expected value per one-and-one situation.

Which One When?

Donna and Adelyn play for their high school basketball team.

When Donna is in a one-and-one situation, she always makes the first shot. But she then gets so excited that she always misses the second shot.

Adelyn, on the other hand, makes her first shot only half the time. But if she does get it, it means she is "in the zone," and she will always make the second shot.

1. Find each player's expected value in a one-and-one situation.

2. Suppose you're the coach of the team. The game is almost over. You can plan your strategy so that the game ends with either Donna or Adelyn in a one-and-one situation.

   Under what circumstances would you prefer to have Donna shooting? When would you prefer Adelyn?

   Explain your reasoning.

3. What does Question 2 tell you about the importance of expected value?

A Fair Dice Game?

In a two-person game, one person throws a pair of dice. If the sum is 2, 11, or 12, you win. If the sum is 7, the other person wins. If the sum is anything else, no one wins, and you throw again. You keep throwing until someone wins.

1. Do you think this game is fair? In other words, are both players equally likely to win? Explain why or why not. Use at least one rug diagram in your explanation.

2. If you think it *is* a fair game, make up another dice game that you think *is not* fair. If you think the game *is not* fair, make up another dice game that you think *is* fair. Once again, use a rug diagram in your explanation.

More Martian Basketball

You may recall from *Martian Basketball* that instead of having one-and-one free throw situations, Martians have one-and-one-and-one situations.

Here is information on three of the Martian Basketball Association All-Star players.

- Splurge Rip is a 70% shooter in any situation.
- Crago Dit makes 70% of first shots, 60% of second shots, and 90% of third shots.
- Lufy Boz makes 80% of first shots, 50% of second shots, and 90% of third shots

Suppose the Martian All-Stars are playing the team from Venus. For the last play of the game, the Martians get to choose one of their players for a one-and-one-and-one situation. Who would be the best person to have shooting? Who would be the worst? How might it depend on the score at that point?

Write an analysis of different cases and how you would decide each one.

Interruptions

Al and Betty are at the park, flipping coins. Al gets 1 point if the coin is heads, and Betty gets 1 point if the coin is tails.

The first one to reach 10 points wins a prize of $15.

With Al leading by a score of 8 to 7, their parents interrupt the game and tell Al and Betty it's time to go home. The friends decide that rather than finish the game another time, they will just give out the prize now.

Al says that because he was leading, he should get the prize. Betty figures that each point should be worth $1, so Al should get $8 and she should get $7.

One of the parents suggests that they should figure out the probability each person had of winning and divide the money according to that.

1. How should they divide the money if they take this advice? Explain your results carefully.

2. Describe two other possible incomplete games. Figure out how Al and Betty should divide the money using the parent's system.

Paying the Carrier

This activity is similar to *The Carrier's Payment Plan Quandary.* Now you are the customer. Instead of paying $5 weekly, you decide to offer your newspaper carrier a different payment method.

The carrier will roll a pair of dice. If the sum is 4 or less, the carrier will get $20. If the sum is more than 4, the carrier will get some other fixed amount.

1. You want this payment method to be equivalent, in the long run, to the $5 per week payment. What should you choose for the other fixed amount—that is, the amount the carrier receives for any sum over 4?

2. Describe how you found the answer to Question 1.

3. Now suppose that the normal weekly cost of the paper is Y, but the carrier still gets $20 for a sum of 4 or less. What should you choose for the other fixed amount the carrier gets (for any sum over 4) in order to make the payoff fair in the long run? You might work with some specific values for Y.

Pig Tails Decision

Pig Tails is a variation on the game of Pig.

> Each turn consists of flipping a coin until either you decide to stop or you get tails. If you stop before getting tails, your score for the turn is the number of heads you got—that is, the number of times you flipped.

Suppose you walk into a room where your friend is in the middle of a game of Pig Tails. Your friend has to leave and asks you to finish the game. Your friend has gotten four heads so far this game.

Do you flip again? Justify your answer.

Get a Head!

At a school fund-raiser, students set up a booth with this game.

> You start flipping a regular coin. Each time you get heads, you get a payoff of $1. If you get tails, the game ends, and you get to keep the money you've won so far. If you get tails on your first flip, you get nothing.
>
> If you get ten straight heads, you get $10, but the game ends and you are given a $50 bonus.

For example, if you flip four heads and then tails, you win $4. If you flip ten heads, you win $60 altogether.

1. Suppose the school charges $2 to play and that each of the 1000 students at the school plays five times. How much profit should the school expect to make?

2. The students are considering eliminating both the limit of ten heads and the bonus. How much profit should they expect to get after making these changes? You may need to give an approximate answer to this problem. Explain your reasoning.

Pig Strategies by Algebra

In *The Best Little Pig* and in *Big Pig Meets Little Pig*, you looked at the question of when it pays for a player of Little Pig or Big Pig to draw or roll again.

In those activities, you assumed that a group of students each had a particular score. You then found the expected value if they each drew or rolled once more. Now you will work more generally.

1. Assume that a player has exactly *S* points in Little Pig.

   a. Find an algebraic expression in terms of *S* for that player's expected value if the player decides to draw one more cube and then stop.

   b. For what values of *S* is your expression greater than *S*? In other words, when does the player gain in the long run by drawing again?

   c. For what values of *S* is your expression less than *S*?

2. Now do the analogous questions for Big Pig. That is, suppose a player has *S* points in Big Pig.

   a. Find an algebraic expression in terms of *T* for that player's expected value if the player decides to roll the die one more time and then stop.

   b. For what values of *T* is that expression greater than *T*? In other words, when does the player gain in the long run by rolling again?

   c. For what values of *T* is your expression less than *T*?

Fast Pig

Fast Pig is another variation of the game of Pig.

Instead of rolling one die again and again, you roll several dice at once, and you get only one roll per turn. That's what makes it fast.

If none of the dice comes up 1, your score is the sum of the dice. If one or more of the dice come up 1, your score for the turn is 0.

Although Fast Pig with just one die isn't very exciting, you may want to think about that game to get some ideas about the questions below.

1. Suppose you play Fast Pig with two dice.

    a. What is the probability that neither die will be a 1?

    b. If you consider only those turns with a nonzero score, what is your expected value for a turn of Fast Pig? You might imagine dice with only five sides, labeled 2 through 6.

    c. Taking into account your answers to parts a and b, what is the expected value all together for a turn of two-dice Fast Pig?

2. Answer the same questions for three-dice Fast Pig.

3. Can you generalize the results? You may want to think about one-die Fast Pig as well.

4. What does the analysis of n-dice Fast Pig tell you about the expected value for certain strategies for Pig?

The Overland Trail

Variables, Graphs, Linear Functions, and Equations

The Overland Trail—Variables, Graphs, Linear Functions, and Equations

A Journey Back in Time

This unit follows the movement of settlers along the Overland Trail from Missouri to California in the 1800s. Their journey of almost 2000 miles usually lasted several months. You will follow them and learn how they had to deal with matters of life and death on their journey, such as illness, weather, and lack of food and water.

To begin, you will create your own pioneer family and make decisions about provisions for the trip. By taking on the role of a pioneer family during the unit, you will make many of the same decisions they had to make—about supplies, routes, and transportation.

As your families travel across the continent, you will encounter some important mathematical ideas, such as graphs, lines of best fit, equations and other uses of variables, and **rate** problems. Just as settlers may have done, you will use mathematics on your migration. You'll also learn interesting history as you encounter some of the real-life pioneers who traveled the overland trails.

Stanley Pinkston sets the stage for the long journey west.

Crossing the Frontier

From *Women's Diaries of the Westward Journey*

Between 1840 and 1870, a quarter of a million Americans crossed the continental United States, some twenty-four hundred miles of it, in one of the great migrations of modern times. They went West to claim free land* in the Oregon and California Territories, and they went West to strike it rich by mining gold and silver.

Men and women knew they were engaged in nothing less than extending American possession of the continent from ocean to ocean . . . . The westward movement was a major transplanting of young families. All the kinfolk who could be gathered assembled to make that hazardous passage together . . . .

The emigrants came from Missouri, Illinois, Iowa, and Indiana, and some all the way from New York and New Hampshire. Most of them had moved to "free land" at least once before, and their parents and grandparents before them had similarly made several removals during their lifetime. These were a class of peasant proprietors. They had owned land before and would own land again. They were young and consumed with boundless confidence, believing the better life tomorrow could be won by the hard work of today . . . .

*While the land was offered "free" to these migrants, it was not land that was free for the taking. It was the home of the indigenous peoples who had been living there for thousands of years.

continued ▶

The journey started in the towns along the Missouri River between St. Joseph and Council Bluffs. These settlements came to be known as the jumping-off places. In the winter months emigrants gathered to join wagon parties and to wait for the arrival of kin. It was an audacious journey through territory that was virtually unknown. Guidebooks promised that the adventure would take no more than three to four months' time—a mere summer's vacation. But the guidebooks were wrong. Often there was no one in a wagon train who really knew what the roads would bring, or if there were any roads at all. Starting when the mud of the roads began to harden in mid-April, the emigrants would discover that the overland passage took every ounce of ingenuity and tenacity they possessed. For many, it would mean six to eight months of grueling travel, in a wagon with no springs, under a canvas that heated up to 110° by midday, through drenching rains and summer storms. It would mean swimming cattle across rivers and living for months at a time in tents.

From *Women's Diaries of the Westward Journey* by Lillian Schlissel. Copyright 1983 by Schocken Books, Inc. Reprinted by permission of Schocken Books, published by Pantheon Books, a division of Random House, Inc.

Just Like Today

Throughout history, people have moved from place to place. In this activity, consider a present-day movement of people and compare it with the movement of people along the Overland Trail.

You can consider many different levels of this topic. For instance, you can look at your own family and a move you or your ancestors made, or you can look at the movement of people from one country to another.

Write about the two movements—the Overland Trail and a more recent move. How are they similar? How are they different? You might comment on why the people are moving in each case and how they get from one place to another.

Overland Trail Families

From the Diary of Catherine Haun

Ada Millington was twelve when her family set out for California. It was a large family; her father, who had five children by his first marriage, was traveling with his second wife and their six children. The youngest was a year and a half old. In addition, there were five young hired hands. And there was Mrs. Millington's sister and brother-in-law, their children, and their hired hands. And there was the brother-in-law's sister, stepfather and mother. The party seemed large and secure . . . .

Our own party consisted of six men and two women. Mr. Haun, my brother Derrick, Mr. Bowen, three young men to act as drivers, a woman cook and myself . . . .

A regulation "prairie schooner" drawn by four oxen and well filled with suitable supplies, with two pack mules following on behind was the equipment of the Kenna family. There were two men, two women, a lad of fifteen years, a daughter thirteen and their half brother six weeks of age. This baby was our mascot and the youngest member of the company . . . .

One family by the name of Lemore consisted of man, wife and two little girls. They had only a large express wagon drawn by four mules and meager, but well chosen, supply of food and feed. A tent was strapped to one side of the wagon, a roll of bedding to the other side, baggage, bundles, pots, pans and bags of horse feed hung on behind; the effect was really grotesque . . . .

Mr. West from Peoria, Ill. had another man, his wife, a son Clay about 20 years of age and his daughter, America, eighteen. Unfortunately Mr. West had gone to the extreme of providing himself with such a heavy wagon and load they were deemed objectionable as fellow argonauts. After disposing of some of their supplies they were allowed to join us . . . .

A mule team from Washington, D.C. was very insufficiently provisioned . . . [by] a Southern gentlemen "unused to work . . . ." They deserted the train at Salt Lake as they could not proceed with their equipment . . . .

Much in contrast to these men were four batchelors Messers Wilson, Goodall, Fifield, and Martin, who had a wagon drawn by four oxen and two milch cows following behind. The latter gave milk all the way to the sink of the Humboldt where they died, having acted as draught animals for several weeks after the oxen had perished. Many a cup of milk was given to the children of the train and the mothers tried in every way possible to express their gratitude.

From *Women's Diaries of the Westward Journey* by Lillian Schlissel. Copyright 1983 by Schocken Books, Inc. Reprinted by permission of Schocken Books, published by Pantheon Books, a division of Random House, Inc.

Creating Families

Your group is responsible for the planning and travel of four family units on the wagon train. In this activity, you will decide on the members of each of your families.

Here are some general conventions.

- Anyone more than 14 years old is considered an adult.
- A *young child* is anyone less than 6 years old.
- An *older child* is anyone from the age of 6 through the age of 14.
- Only adults have children.
- Hired hands are adults and are not related to the other members of the family unit. They are considered to be part of the family with which they travel.
- A wagon can accommodate at most 6 people. A family unit of 6 or fewer people needs only one wagon, but a family unit of between 7 and 12 people needs two wagons, a family unit of between 13 and 18 people needs three wagons, and so on.

Four categories can be used to describe types of family units that traveled on the trail. Each family type is subject to certain **constraints.** That is, there are conditions or criteria that define the type of family unit.

Your group will complete these steps.

- Create one family of each type. Each group member will have final responsibility for one family.
- Make a complete list of all the people in each family. Give each person a first and last name, an age, and a gender. Use names from the list in *Overland Trail Names*.
- Record how many adult men, adult women, older children, and young children there are in each family. You will use this information in later activities.

continued ▶

The Small Family

Some family units were quite small by the standards of the time. A *small family* fits these constraints.

- There are more adults than children.
- There is at least one child.
- There are at least as many adult men as adult women.
- There is at most one married couple.
- There is at least one pair of adult siblings.
- The number of people in the family unit is less than eight.

The Large Family

In addition to the father and mother, a *large family* has many children and many other adults. A *large family* meets these constraints.

- The number of children is greater than the number of adults.
- There are between one and six hired hands.
- Four generations of family members are represented.
- At least two married couples are present in the group.
- There are more young children than there are older children.
- There are at most 25 family members.

continued ▶

The Nonfamily

Some groups traveling on the trail were hardly families at all. A *nonfamily* is one that fits these constraints.

- There are no married couples.
- Any adult women in the group are traveling with a brother or a father.
- There is at most one adult woman for every four adult men.
- The number of adult men in the group can range from two to eight.
- There is no more than one child for every five adults.
- There are at least as many older children as young children.

The Conglomerate Family

Sometimes small families banded together into a single unit for the trip. Such a *conglomerate family* meets these constraints.

- There are two or three partial families in the unit.
- Each partial family includes at least one adult.
- Each partial family has fewer than five people in it.
- The total number of children equals at least one-third the total number of adults.
- There are more young children than older children.

Overland Trail Names

The names in these lists are taken from sources contemporary to the period of the Overland Trail.

Last names

Ackley	Cazneau	Frizzell	Kelsey	Smith
Adams	Clappe	Frost	Ketcham	Spencer
Agatz	Clarke	Fulkerth	Knight	Stewart
Allen	Collins	Geer	Mason	Stone
Ashley	Colt	Goltra	Millington	Tabor
Bailey	Cooke	Hall	Minto	Ward
Ballou	Cox	Hanna	Norton	Washburn
Behrins	Dalton	Haun	Parker	Waters
Bell	Deady	Helmick	Parrish	Welch
Belshaw	Duniway	Hines	Pengra	Whitman
Bennett	Findley	Hixon	Porter	Wilson
Bogart	Fish	Hockensmith	Powers	Wood
Brown	Foster	Hodder	Pringle	
Buck	Fowler	Hunt	Rudd	
Butler	French	Jones	Sanford	
Carpenter	Frink	Kellogg	Sawyer	

Female first names

Abigail	Catherine	Esther	Louise	Nancy
Ada	Celinda	Hallie	Lucinda	Rebecca
Amelia	Charlotte	Helen	Lucy	Roxana
America	Clara	Jane	Lydia	Sarah
Ann	Elizabeth	Julia	Margaret	Susan
Caroline	Ellen	Lavinia	Mary	Velina

Male first names

Addison	Evan	Henry	Lafayette	Robert
Alpheus	Ezra	Holmes	Lewis	Samuel
Charles	Francis	James	Moses	Solomon
Dexter	George	Jared	Perry	Thomas
Edward	Gilbert	Jay	Peter	Tosten
Enoch	Godfry	John	Richard	William

The Haybaler Problem

Imagine you have five bales of hay. For some reason, instead of being weighed individually, they were weighed in all possible combinations of two: bales 1 and 2, bales 1 and 3, bales 1 and 4, bales 1 and 5, bales 2 and 3, bales 2 and 4, and so on.

The weight of each combination was written down *without keeping track of which weight matched which pair of bales.* The weights in kilograms were 80, 82, 83, 84, 85, 86, 87, 88, 90, and 91.

○ Your Task

Your initial task is to find out how much each bale weighs. Is your answer the only possible set of weights? Explain how you know.

Once you are done looking for solutions, look back over the problem to see if you can find some easier or more efficient way to find the weights.

○ Write-up

1. *Problem Statement*

2. *Process:* The process is especially important in this problem. Include a description of any materials you used. Be sure to discuss all the ways you tried to attack the problem, even though they all didn't lead to the correct answer.

   Also discuss any insights you had about other ways you might have solved it.

3. *Solution:* Show both how you know your weights work and how you know that you have not missed some other possibilities.

4. *Extensions*

5. *Self-assessment*

Hats for the Families

Everyone on the Overland Trail will be spending long hours in the hot sun. Before setting out, each person will need one good hat to help keep the sun off.

1. What is the minimum and maximum number of hats that might be needed for each type of family unit?

2. What would be the minimum and maximum number of hats that might be needed for a wagon train consisting of one family unit of each type?

3. Estimate the number of hats that might be needed for all the family units in your class. Explain your reasoning.

Family Constraints

The families in this activity do not necessarily belong to an Overland Trail wagon train.

1. The Hickson household contains three people of different generations. The total of the ages of the three family members is 90.

   a. Find reasonable ages for the three Hicksons.

   b. Find another reasonable set of ages for them.

   c. One student, in solving this problem, wrote

   $$C + (C + 20) + (C + 40) = 90$$

   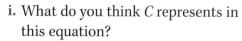

      i. What do you think C represents in this equation?

      ii. What do you think the student had in mind in using the numbers 20 and 40?

      iii. What set of ages do you think the student came up with?

2. There are four members in the Jackson family, again representing three generations. As in the Hickson household, the total of the ages of these people is 90.

   a. Find a reasonable set of ages in which there are two children, one parent, and one grandparent.

   b. Find a reasonable set of ages in which there is one child, two parents, and one grandparent.

   c. Find a reasonable set of ages in which there is one child, one parent, and two grandparents.

continued ▶

d. In working on Question 2a, one student wrote

$$C + C + (C + 18) + (C + 36) = 90$$

    i. What does C represent here?

    ii. What assumption seems to be implied by the fact that the equation begins with $C + C$?

    iii. What do you think the student had in mind in using the numbers 18 and 36?

    iv. Why do you think this student used 18 and 36 while the student in Question 1c used 20 and 40?

    v. What set of ages do you think this student came up with?

3. Clara Dalton is the mother of Lucinda, one of a set of quadruplets. Lucinda is the mother of triplets.

The family tree here shows Clara, her quadruplets, and Lucinda's triplets.

The sum of the ages of these eight people is 201.

Find a reasonable set of ages for these people.

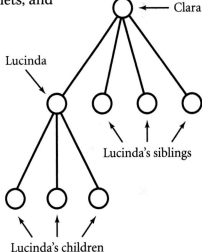

Planning for the Long Journey

The first leg of the journey takes your families from Westport, Missouri, to Fort Laramie, Wyoming, some 600 miles away. Wagon trains traveled about 20 miles each day during this part of the journey.

In this activity, you will choose supplies for the first leg of the journey.

Part I: Generating Ideas

1. Identify those supplies you think the four Overland Trail families will need. Brainstorm to create a list. Be specific. Don't just say tools—list the tools. Don't just say food—name the different kinds of food the family will need.

2. As you brainstorm and compile this tentative list, questions may occur to you that you cannot answer. Write down these questions to share with the rest of the class.

Part II: Making Decisions

3. *Overland Trail Price List* gives the cost for certain items that you may wish to purchase. Assume that you have $10 to spend on supplies for each Overland Trail person. Decide how much of each item you want to buy for your group's Overland Trail families.

   Your purchase list must include gunpowder, sugar, and beans. You may wish to save some money for use along the trail.

Overland Trail Price List

Item	Cost
flour	2 ¢/lb
biscuits	3 ¢/lb
bacon	5 ¢/lb
coffee	7 ¢/lb
tea	50 ¢/lb
sugar	10 ¢/lb
lard	6 ¢/lb
beans	8 ¢/lb
dried fruit	24 ¢/lb
salt	4 ¢/lb
pepper	4 ¢/lb
baking soda	4 ¢/lb
gunpowder	25 ¢/lb

Prices in the list are taken from *Women and Men on the Overland Trail* by John Mack Faragher (New Haven, CT: Yale University Press, 1979). Copyright 1979 by Yale University Press.

The Search for Dry Trails

Families used to arrive in the Westport area (near present-day Kansas City, Missouri) during the winter. Around April, the muddy trails would begin to dry, and the settlers would start their westward journey. But rainy weather could still make roads muddy and slow a wagon train's progress.

The three main trails to the eastern slope of the Rocky Mountains were

- the Smoky Hill Trail, which followed the Smoky Hill River to Denver, Colorado
- the Santa Fe Trail, which followed the Arkansas River to southern Colorado
- the Oregon Trail, which followed the North Platte River to Laramie, Wyoming

Note: These descriptions use the names given by settlers to places and geographical features.

Your Task

Over the years, the owner of the Westport Trading Post has heard from various friends concerning the weather they found on their routes.

The owner has compiled a table, showing the number of rainy days each friend encountered. Different entries for a given trail represent different years on that trail.

continued ▶

Santa Fe Trail		Smoky Hill Trail		Oregon Trail	
Friend's name	Number of rainy days	Friend's name	Number of rainy days	Friend's name	Number of rainy days
William	24	Tosten	18	Enoch	42
Amelia	3	Roxana	16	Sarah	9
Ezra	21	Hallie	13	Godfry	11
Lavinia	5	Ada	14	Alpheus	10
Moses	23	Dexter	19	Ann	12
				Jared	13

A big argument occurs one evening. It seems that family members looking at the same data cannot agree on which trail would be the driest. Of course, the dryness of a trail was not the only factor in choosing the route.

1. If you were on the Overland Trail, which trail would you choose? Why?

2. Addison Pengra says he would choose the trail that had the fewest rainy days on average.

   a. Which trail would Addison choose? Explain your thinking.

   b. What are some advantages and disadvantages of Addison's method?

3. Lydia Pengra says she would make her choice this way: She'd look at the numbers for each trail, put them in order, and then find the number in the middle of the list. Then she'd choose the trail that had the smallest middle number.

   a. Which trail do you think Lydia would choose, based on her method? Explain your thinking.

   b. What are some advantages and disadvantages of Lydia's method?

Setting Out with Variables

You've formed your Overland Trail families, gotten your supplies together, and started off on the journey from Westport, Missouri, toward Fort Laramie, Wyoming. The journey becomes more difficult. Everything changes as the trip progresses—supplies run out and need to be replenished, people fall sick and sometimes die, families decide to turn back, weather and terrain cause the wagon train to alter course. You will see how variables, equations, and algebraic expressions may have helped the settlers plan their journey and meet its challenges.

You will track the progress of your families during a series of activities tied to places and events along the Overland Trail. Variables, expressions, and equations will help you along the way.

To focus on the meaning of variables, students are creating their own "ox expressions" to share with the class.

Shoelaces

Shoelaces are one small item that must be taken on the Overland Trail. In this activity, you will consider how much of this commodity is needed.

Assume that everyone's shoes and boots already have laces. You want to be able to replace each lace once during the journey. Also assume that each pair of shoes or boots needs its own laces.

Here is some detailed information about shoelace requirements.

- Each man needs to bring two pairs of boots and one pair of shoes.
- Each woman needs to bring one pair of boots and two pairs of shoes.
- Each child needs to have three pairs of boots.
- A shoelace for each adult boot is 48 inches long.
- A shoelace for each adult shoe is 32 inches long.
- A shoelace for each child's boot is 24 inches long.

1. How many inches of shoelace does a woman need?

2. How many inches of shoelace does a man need?

3. How many inches of shoelace does a child need?

4. Find the total length of shoelace needed for the specific Overland Trail family for which you are responsible.

5. Describe in words how you used your answers from Questions 1, 2, and 3 to get your answer to Question 4.

Laced Travelers

In this activity, you are told how much shoelace each person needed. These amounts are different from those in *Shoelaces*. Use these new amounts.

Suppose these statements were true in 1852.

- Shoelaces cost 2 ¢ per yard.
- The average number of families in a wagon train was 25.
- The average family had six people in it (counting unmarried relatives and hired hands): two men, one woman, and three children.
- Approximately 150 wagon trains went through Westport, Missouri, in the year 1852 on their way west.
- Each man needed 5 yards of shoelace.
- Each woman needed 4 yards of shoelace.
- Each child needed 3 yards of shoelace.

1. How many yards of shoelace did the settlers who went through Westport in 1852 need altogether?

   Once you've found an answer, describe in words how you did the computation.

2. Write two more interesting questions related to the journey that you can answer from the data for 1852.

3. Answer one of the questions you made up in Question 2. Describe in words how you did the computations.

4. Suppose that in 1853, smaller families were migrating west. The average family size was only five people (one less child). Answer Question 1 for the year 1853, assuming that the other information is unchanged.

To Kearny by Equation

When the first emigrants went west, crossing rivers was dangerous and time-consuming. Travelers were grateful and travel time was shortened when people started ferries to shuttle wagons across the rivers.

The first major stop along the way from Westport to Fort Laramie was at Fort Kearny (now Kearney, Nebraska).

Joseph and Louis Papan, two brothers, were of mixed blood. They had one Native American parent and one parent of European origin. They operated a ferry over the Kansas River at Topeka, on the way from Westport to Fort Kearny.

We don't know how the Papans set their rates, but you can make these assumptions.

- The fee for crossing the 230-yard-wide river was $2 for each wagon.
- The ferry captain received pay of 30¢ per hour from the Papans for the time he spent going back and forth.

continued ▶

The Papan brothers could then calculate the profit they made using the equation

$$\text{profit} = 2W - 0.3H$$

Here, W is the number of wagons that crossed the river, and H is the number of hours that the ferry captain spent going back and forth. This profit formula takes into account the captain's salary, but it does not take into account the Papans' other expenses, such as upkeep of the boat.

1. Explain why this formula makes sense.

2. Write an expression for the amount that *each* Papan brother made. Assume that they split the profit equally.

Now add these assumptions.

- A round trip on the ferry took 20 minutes.
- A wagon could hold at most 6 people. A group of between 7 and 12 people required two wagons, a group of between 13 and 18 required three wagons, and so on. Every group had at least one wagon.
- The ferry could hold up to four wagons and their passengers. Larger families required more than one ferry trip.

3. How much profit would each Papan brother make from the family unit for which you are responsible?

4. How much profit would each Papan brother make from your group's four family units?

The Vermillion Crossing

Louis Vieux was a business manager, interpreter, and chief of the Potawatomi. He made many trips to Washington to consult with officials about Native American affairs.

Vieux was also a ferry operator. He operated a ferry and toll bridge over the Vermillion River, the third major river crossing in Kansas.

Vieux charged a certain amount for each wagon plus an additional amount for each person, with different amounts for men, women, and children. Suppose that the amount he charged was given by the equation

$$\text{price to cross (in dollars)} = 0.5W + 0.25M + 0.1F + 0.05C$$

Here, W represents the number of wagons, M the number of men, F the number of women, and C the number of children.

1. Use this formula to explain what Vieux charged in each cost category (that is, for each wagon, for each man, and so forth).

2. What would be the crossing cost for the family unit for which you are responsible? As in *To Kearny by Equation*, assume that a wagon could hold at most 6 people—so a group of between 7 and 12 people required two wagons, a group of between 13 and 18 required three wagons, and so on.

3. What would be the total cost for your group's four family units?

Ox Expressions

The table defines some symbols as variables to represent certain quantities. For example, *F* stands for "the number of **F**amilies in a wagon train." The **boldface** letters will remind you of what each symbol represents. A specific numeric value is provided for each variable. Treat this value as constant for all cases. For example, assume that *every* wagon train contains 25 families. (These values will probably not be the actual numbers in your class wagon train.)

Symbol	Meaning	Numeric value
F	the number of **F**amilies in a wagon train	25 families per train
M	the number of **M**en in a family	2 men per family
W	the number of **W**omen in a family	1 woman per family
C	the number of **C**hildren in a family	3 children per family
V	the number of wagons (**V**ehicles) per family	1 wagon per family
T	the number of wagon **T**rains in one year	150 trains per year
Y	the number of pairs (**Y**okes) of oxen per wagon	3 yokes per wagon
A	the number of oxen (**A**nimals) per yoke	2 oxen per yoke
P	the weight of one ox (in **P**ounds)	1200 pounds per ox
L	the **L**oad for one wagon (in pounds)	2500 pounds per wagon
G	the amount of **G**rass eaten by one ox in one day (in pounds)	40 pounds of grass per ox per day
H	the amount of water (**H**$_2$O) consumed by one ox in one day (in gallons)	2 gallons of water per ox per day
B	the amount of water (**B**everage) consumed by one person in one day (in gallons)	0.5 gallons of water per person per day
D	the number of **D**ays on the trail	169 days

continued ▶

Using the given letters, it is possible to write many different algebraic expressions. Although you can always **substitute** numbers for the letters and do the arithmetic, most of the expressions have no real meaning.

For example, for the expression *MG*, you can multiply the number of men per family by the amount of grass an ox can eat in a day. However, the product you get doesn't have any useful application. In other words, *MG* doesn't really mean anything.

Some expressions *do* have a meaning. For example, *FC*, the number of families in a wagon train times the number of children in a family, represents the total number of children traveling in a wagon train. So the expression *FC* has meaning.

The phrase "the number of children traveling in the train" is a concise way to describe the number represented by *FC*. We will call this the **summary phrase.**

The table tells you that there are 25 families in a wagon train, so $F = 25$, and that there are 3 children in a family, so $C = 3$. Therefore, $FC = 25 \cdot 3 = 75$; there are 75 children in a wagon train. Even if the numbers were different, *FC* would still represent the number of children in a wagon train.

Your Task

Your task is to come up with as many meaningful algebraic expressions as you can, using the symbols in the table. For each expression, go through these steps.

- Write the expression.
- Explain what the expression means, using a summary phrase.
- Give the numerical value of the expression, based on the values of the individual variables given in the table.

Ox Expressions at Home

In this activity, you continue to work with algebraic expressions and summary phrases.

You will be given specific algebraic expressions and asked to write summary phrases for them. You also will be given specific summary phrases and asked to write algebraic expressions for them.

Reminder: The summary phrase for *FC* is "the number of children in the wagon train" and not "the number of families in a wagon train times the number of children in a family."

The symbols below are the same as those used in *Ox Expressions.* Though no specific numeric values are assigned, you should assume that each symbol represents a single number.

Symbol	Meaning
F	the number of **F**amilies in a wagon train
M	the number of **M**en in a family
W	the number of **W**omen in a family
C	the number of **C**hildren in a family
V	the number of wagons (**V**ehicles) per family
T	the number of wagon **T**rains in one year
Y	the number of pairs (**Y**okes) of oxen per wagon
A	the number of oxen (**A**nimals) per yoke
P	the weight of one ox (in **P**ounds)
L	the **L**oad for one wagon (in pounds)
G	the amount of **G**rass eaten by one ox in one day (in pounds)
H	the amount of water (**H**$_2$O) consumed by one ox in one day (in gallons)
B	the amount of water (**B**everage) consumed by one person in one day (in gallons)
D	the number of **D**ays on the trail

continued ▶

1. Write a summary phrase for the expression $W + M + C$.

2. Write an algebraic expression for the water consumed in a day by a family.

3. Write a summary phrase for the expression $D(H + B)$.

4. Write an algebraic expression for the number of people in a wagon train.

5. Write a summary phrase for the expression FM.

6. Write an algebraic expression for the amount of water consumed by an ox on the trip.

7. Does the expression WL have a meaning? If so, what is it?

8. Make up a meaningful algebraic expression of your own and give a summary phrase for it.

If I Could See This Thing

No nation was safe from the ravages of smallpox, cholera, measles, scarlet fever, influenza, and tuberculosis. These diseases, which were imported from Europe, took a great toll on Native Americans. Disease brought death, destruction, and untold misery, killing more people than warfare, slavery, or starvation.

This passage is from a description by George Bent of the Southern Cheyenne nation.

> In '49, the emigrants brought cholera up the Platte Valley, and from the emigrant trains it spread to the Indian camps. "Cramps" the Indians called it .... On the Platte whole camps could be seen deserted with tepees full of dead bodies .... Our Tribe suffered very heavy loss; half of the tribe died, some old people say.
>
> My Grandmother took the children that summer ... to the Canadian [to get] medicine. During the medicine dance an Osage visitor fell down in the crowd with cholera cramps. The Indians broke camp at once and fled in every direction. Here a brave man ... mounted his horse ... and rode through camp shouting, "If I could see this thing, if I knew where it was, I would go there and kill it." He was taken with cramps as he rode.

From *Life of George Bent* by Savoie Lotinville (Norman: University of Oklahoma Press, 1968).

continued ▶

1. Some historians estimate that between 1492 and 1900, the Native American population decreased by about 90%.

   Use variables and an equation to show how you would find the population of Native Americans at the end of this period if you knew their population at the beginning.

   Suggestion: Pick a number that you think might represent the population in 1492, and figure out what the 1900 population would have been. Then describe your computation in words. Finally, put the relationship into equation form using variables.

2. Death occurred among travelers on the Overland Trail as well.

   The fatality rate differed from one wagon train to the next. For this activity, assume that five percent of the adults and ten percent of the children died of cholera on the road from Fort Kearny, Nebraska, to Fort Laramie, Wyoming.

   a. Figure out how this will change the size of your total class wagon train.

   b. Put the result from Question 2a into an In-Out table with two inputs—the number of adults and the number of children. The output is the total number of people who will be left in a wagon train after this leg of the journey.

      Add two more rows to this table. Use your own choice of values for the number of adults and number of children. For each new row, calculate the number of people in the wagon train at Fort Laramie.

   c. Introduce variables and write a rule for the In-Out table in Question 2b.

The Graph Tells a Story

They say that a picture is worth a thousand words. While you rest in Fort Laramie and prepare for the next leg of your journey, you look at some posters describing certain aspects of what lies ahead. These posters contain graphs that show how two quantities are related. A graph can show the connection between days on the trail and dwindling supplies, or days elapsed and distance from a destination.

As you will see, graphs are closely related to equations and In-Out tables. As you imagine yourself on the Overland Trail, you will want to do just like the real-life travelers. You will gather as much information as you can and make sense of it in whatever way best helps you to arrive safely.

Not everyone in the nineteenth century traveled to the West in ox-drawn wagons. Some made the journey by sea. The new POW in this segment, *Around the Horn,* poses a question that might have been on the minds of some travelers on board a ship.

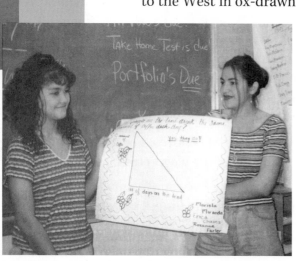

Mariela Miranda and Erica Chavez present their graph to the class.

Wagon Train Sketches and Situations

A graph sketch can be used to describe a real situation. For example, this graph sketch shows that the number of pairs of shoes and boots needed for a wagon train family unit depends on the number of people in the family unit.

Your task will be to look at sketches and say what you think is happening. You will also create sketches to represent situations.

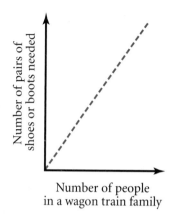

Number of people
in a wagon train family

Part I: From Sketch to Situation

When you arrive in Fort Laramie, you see posters promoting the journey westward. These posters contain graph sketches describing relationships concerning the trip west.

For each graph, describe what's happening in the situation that the graph represents. Then answer the specific question for that graph.

1. According to this graph sketch, do people on the trail drink the same amount of coffee each day? Explain your answer.

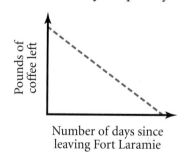

Number of days since
leaving Fort Laramie

2. What do you think is happening at the points on the graph sketch labeled *A*, *B*, and *C*?

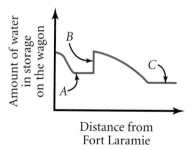

Distance from
Fort Laramie

continued ▶

3. At what part of this graph sketch was the wagon train moving fastest? You can trace the graph onto your own paper and mark your answer.

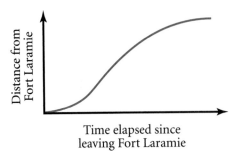

Time elapsed since
leaving Fort Laramie

4. Why does this graph sketch consist of individual dots instead of a line?

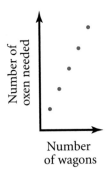

Number
of wagons

Part II: From Situation to Sketch

Here are several descriptions of relationships. For each description, sketch a graph that illustrates the relationship.

5. Wagons of the same size and type can hold a fixed number of people. Make a graph sketch that shows the relationship between the number of wagons (of a fixed size and type) and the number of people those wagons can carry.

6. As the number of settlers on the trail increased, the buffalo population declined. Make a graph sketch that shows the relationship between the number of settlers and the buffalo population.

7. A man rides his horse from his ranch to a neighboring ranch without stopping. His route takes him through the center of town. Make a graph sketch that shows the relationship between how long he has been riding and his distance from the center of town.

Graph Sketches

Part I: Sketches to Situations

Each graph sketch illustrates a relationship between two quantities.
In each case, describe a situation that is illustrated by the graph.

1.

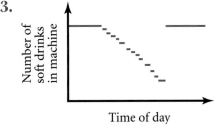

2.

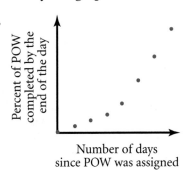

3.

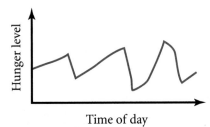

4.

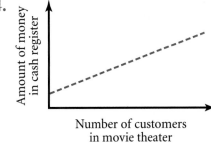

Part II: Situations to Sketches

Describe a situation with a possible relationship between two
quantities.

Write this description on a separate piece of paper. On the back,
sketch the appropriate graph for that relationship. Remember to
label the axes.

In Need of Numbers

Graph sketches describe a situation, but the description is more complete when the graph includes numeric information.

You can provide this information by putting a scale on each **axis**. The scale shows the numeric values that the points on each axis represent.

To scale an axis, decide what range of values is appropriate for the situation and for the quantities involved. You also have to decide how to display the scale on each axis.

For each sketch, go through these three steps.

* Make a copy of the sketch on graph paper.
* On your copy, scale the axes with appropriate values.
* Write down why your scales are reasonable and the assumptions you made.

1.

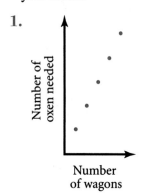

2.

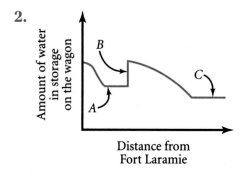

continued ▶

3.

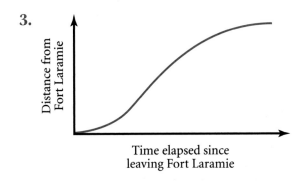

4.

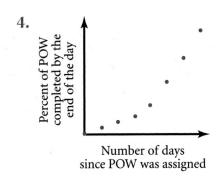

5.

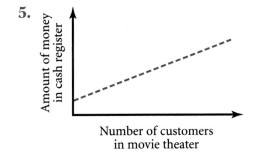

The Issues Involved

1. In *In Need of Numbers,* you put appropriate scales on the axes of different graphs.

   Make a list of difficulties you had and questions you would like answered that are related to scaling the axes of a graph.

2. Here are some questions about scaling. Use examples to explain your thinking. Be detailed in your explanations.

   a. Should the **vertical** axis always begin at zero? What is the effect if the axis does not begin at zero? What about the **horizontal** axis?

   b. Once you have decided on a scale, how do you decide which numbers to write along the axes?

3. The graph shows the average height of boys in the United States at different ages.

   The graph seems to suggest that boys grow at a constant rate through age 5.

   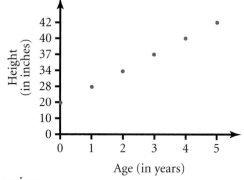

   a. Why might someone make this conclusion from a quick glance at the graph?

   b. Why is this conclusion incorrect?

   c. Redraw the graph so it is not misleading.

4. Suppose you wanted to sketch a graph showing the number of livestock deaths during the Overland Trail trip. Would you use a **continuous** or a **discrete graph** to represent this situation? Why?

Out Numbered

The scaled graphs in this activity are similar to examples you have seen before. Base your answers to the questions *on the scales shown in these graphs.*

1. This graph shows the number of people that can be carried in a given number of wagons.

   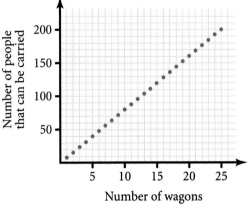

   a. How many people can three wagons carry?

   b. How many people can five wagons carry?

   c. How many people can ten wagons carry?

   d. Make an In-Out table with the information from Questions 1a through 1c. *In* = number of wagons; *Out* = number of people that can be carried.

   e. Find a rule for the number of people that *x* wagons can carry. Use the graph to generate additional rows for the In-Out table if you need more information.

   f. How many people can each wagon carry? Describe, in writing, how this number connects to the rule you found in part e.

2. The next graph shows the amount of coffee left in terms of the number of days since leaving Fort Laramie.

   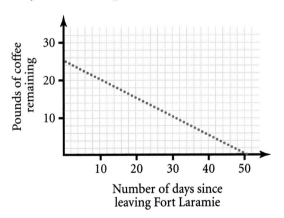

   a. How much coffee was left 10 days after leaving Fort Laramie?

   b. How much coffee was left 15 days after leaving Fort Laramie?

   c. How much coffee was left 35 days after leaving Fort Laramie?

continued ▶

d. Make an In-Out table with the information from Questions 2a through 2c. *In* = number of days since leaving Fort Laramie; *Out* = number of pounds of coffee left.

e. Find a rule for the amount of coffee left *x* days after leaving Fort Laramie. Add rows to the In-Out table if you need more information.

f. How much coffee was there when the group left Fort Laramie? How much did they use each day? Describe, in writing, how these amounts connect to the rule you found in part e.

3. The next graph shows the amount of money in a movie theater cash register as a function of the number of customers in the theater.

a. How much money would be in the cash register if there were 25 customers?

b. How much money would be in the cash register if there were 75 customers?

c. How much money would be in the cash register if there were 250 customers?

d. Make an In-Out table with the information from Questions 3a through 3c. *In* = number of customers; *Out* = amount of money in cash register.

e. Find a rule for the amount of money in the cash register if there were *x* customers. Add rows to the In-Out table if you need more information.

f. How much money would be in the cash register if there were no customers? How much would the amount of money change for every additional customer? Describe, in writing, how these amounts connect to the rule you found in part e.

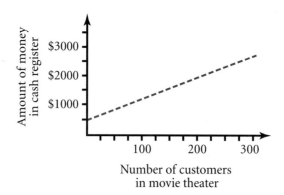

From Rules to Graphs

In *Out Numbered,* you started from graphs, made In-Out tables, and then found rules for those tables.

This process can be reversed. You can start from a rule, make an In-Out table by finding pairs of numbers that fit the rule, and then create a graph using the pairs in the table to give you points on the graph.

You have come across many different In-Out rules so far. Sometimes the rules came from problem situations. Other rules came from tables that had no context. In either case, the rule itself can be used to create a graph. Do not restrict yourself to the first **quadrant** or to whole numbers. Consider all numbers, including negative and noninteger values.

For Questions 1 through 3, do these steps.

    a. Make a table with some In-Out number pairs that fit the rule.

    b. Plot the number pairs (called **ordered pairs**) from your table on a **coordinate system.** Use appropriate scales.

    c. Continue until you have a good idea of what the whole graph looks like. Then sketch the graph.

1. *Out* $= 4 \cdot In - 4$

2. *Out* $= In^2$

3. *Out* $= 550 - 20 \cdot In$

Often the standard letters x and y are used instead of the words *In* and *Out.* When you see an equation using x and y, you should assume that x represents the **independent variable,** or the *In,* which goes on the horizontal axis. The y represents the **dependent variable,** or the *Out,* which goes on the vertical axis.

4. Graph each equation, using a complete coordinate system.

    a. $y = 5x + 3$

    b. $y = 10 - 2x^2$

Around the Horn

Instead of going overland to reach California, some families took a ship around Cape Horn at the tip of South America.

Suppose a ship leaves New York for San Francisco on the first of every month at noon. At the same time, a ship leaves San Francisco for New York.

Suppose also that each ship arrives exactly 6 months after it leaves.

If you were on a ship leaving from New York, how many ships from San Francisco would you meet?

○ *Write-up*

1. *Problem Statement:* If there were any assumptions that you needed to make to do this problem, be sure to state them clearly.

2. *Process:* Include any diagrams or descriptions of materials you used in working on this problem.

3. *Solution*

4. *Extensions*

5. *Self-assessment*

You're the Storyteller: From Rules to Situations

In *Family Constraints,* you used equations containing variables to help you understand some questions about the ages of different people. In *Ox Expressions* and *Ox Expressions at Home,* you used variables based on the Overland Trail context to create meaningful expressions by combining variables.

In this activity, you will work with equations, variables, and situations in another way. Your goal is to create a situation and question that match each equation.

Do steps a and b for Questions 1 through 5.

    a. Create a situation and a question about that situation so that solving the equation will give you the answer to your question. State clearly what the variable in the equation represents.

    b. Find the number that makes the equation true and explain what that number means in your situation.

  1. $4a = 12$

  2. $r + 5 = 20$

  3. $2m + 1 = 11$

  4. $\frac{t}{3} = 8$

  5. $13 - f = 6$

Traveling at a Constant Rate

Graphs don't just tell stories; they can also be very useful in making predictions. The lives of travelers on the Overland Trail often depended on their ability to foresee accurately what could happen to them. The data they worked with didn't often fit neatly into simple rules, so they had to make approximations.

As travelers set out from Fort Laramie toward Fort Hall, Idaho, one of the decisions they made involved a shortcut called Sublette's Cutoff. When you get to that activity, think about what choice you might have made.

Your experience up to this point on the journey has laid the groundwork for understanding that relationships can be represented in four interrelated ways—as situations, graphs, tables, and rules. Fortunately, you have modern technology, which was not available to help nineteenth-century emigrants see the connections among these mathematical representations.

Rodrick Rogers creates a graph based on data.

Previous Travelers

The first settlers had to make the long journey west without advice from previous travelers. However, later wagon trains used information from the early travelers to decide on the supplies to buy for their journey.

While in Fort Laramie, Wyoming, you get a letter from friends describing the supplies they and others used on the leg of the trip from Fort Laramie to Fort Hall, Idaho. Use the information in the letter to answer these questions.

1. Make a graph displaying all the data relating the number of people and the amount of beans they used. Use appropriate scales for the axes. Then make a similar graph for sugar and another for gunpowder.

2. Do these steps for each graph in Question 1.

    a. Sketch what you consider to be the **line of best fit** for the graph; that is, find the straight line that you think best fits the data.

    b. Make an In-Out table *from your line*. Determine a rule that fits your table or that comes reasonably close.

    c. Use either the In-Out table or your graph to find the quantity of each item needed for each of your group's four family units.

continued ◗

Dear friends,

We've arrived in Fort Hall after many adventures, both good and bad. It would take me forever to describe all that happened. We can talk about all of that when we meet up again in California.

I know that you're anxious for some practical information for your own trip. Several of the families on our wagon train kept track of the quantities of various goods they actually needed on the journey. The families were of different sizes, so this information should help you and your friends decide how the amounts vary from group to group. I know this won't answer all your questions, but it's a start.

Number of people	Pounds of beans	Pounds of sugar	Pounds of gunpowder
5	61	20	3
8	95	50	4
6	56	30	2.5
7	75	23	4.1
11	125	60	5
10	135	40	5.8
5	80	39	1.8
7	100	44	3.8
10	103	53	4.3
6	75	35	3.6
8	100	35	3.2
7	105	36	3.1
9	125	45	4.7
12	150	55	6.1
10	125	31	5.2

Well, good luck to you all!

The Helmicks

Broken Promises

Over the years, Native Americans were forced onto smaller and smaller parcels of land. Though the U.S. government signed treaties with the native peoples, the government repeatedly broke those treaties.

1492

The map labeled 1492 shows the outline of the contiguous 48 states of the United States. In each of the remaining maps, the dark portion represents that part of the land remaining to Native Americans in the given year.

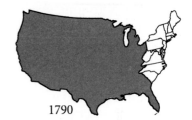

1790

1. The total area shown for 1492 is approximately 3,000,000 square miles. Using that estimate, approximate the area of Native American land at each of the given times. *Suggestion:* You may want to trace each map onto grid paper and use one square as the unit of area.

1830

2. Make a graph showing the relationship between the passage of time and the area of Native American land. Be careful to use appropriate spacing on your time scale.

3. If it were 1861 and you were only looking at the part of your graph showing what had happened up to that year, what would you predict as the area of Native American land in the year 2020?

1860

4. Based on the graph, what prediction might you have made in 1900 about Native American land in the year 2020?

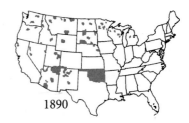

1890

Maps reproduced by permission of Thunderbird Enterprises, Phoenix, Arizona.

Sublette's Cutoff

As more and more emigrants made the journey west, scouts found shortcuts to lessen the travel time. Help provided by Native Americans in the area was also important in facilitating the journey.

The Shoshone, Assiniboin, and Crow nations were prominent in the area of what is now Wyoming and Idaho.

One shortcut between Fort Laramie and Fort Hall was known as Sublette's Cutoff. The cutoff began just past South Pass in Wyoming, about 250 miles west of Fort Laramie. It ended near the Wyoming-Idaho border.

This shortcut saved 50 miles and a week of travel, but it crossed a dry and barren stretch of land. It was a grueling route of 15 days with little grass and no water.

Three families decide to attempt to cross Sublette's Cutoff. The table shows how much water each family has left at the end of the second, fifth, and ninth days.

Gallons of Water Remaining

Family	Day 2	Day 5	Day 9
Jones	49	40	26
Sanford	68	48	25
Minto	30	21	14

1. Graph the water supply data for all three families on the same set of axes. Colored pens or pencils might help.

2. Based on this information, who do you think will make it, and who will not? Explain your reasoning.

3. Is there a time when all three families will have about the same amount of water left? If so, when?

4. Estimate how much water each family used per day and how much water each family started with.

Who Will Make It?

Dagny Appel bought an almanac at the trading post in Fort Laramie. The almanac included predictions about the weather, crops, and livestock.

Dagny was a tireless planner. She shared the predictions with almost everyone in her wagon train.

One particular prediction worried Dagny: The Green River, some 330 miles away, was expected to flood in 30 days, which was about how long it might take to get there.

Three wagon trains kept track of their distances remaining from the Green River. The table shows how far each wagon train was from the river at the end of three different days.

Distance to the River (in miles)

Wagon train	Day 4	Day 7	Day 11
Fowler	270	235	185
Belshaw	285	260	230
Clappe	280	245	200

1. Graph the data for all three wagon trains *on the same set of axes.* Colored pens or pencils might help. Sketch your line of best fit for each family's data.

2. If the almanac is correct about when the flood will take place, who will make it to the Green River before the flood, and who will not? Explain your reasoning.

3. If the almanac is wrong and all three groups make it past the river, which wagon train do you think will arrive at the river first? Last? Explain your reasoning.

4. When the first of these wagon trains gets to the Green River, how far back is the next wagon train? What about the last wagon train? Explain how you found your answers.

5. Estimate the distance covered per day for each wagon train.

The Basic Student Budget

Cal, Bernie, and Doc are college students on budgets.

Sometimes the three have a little difficulty keeping to their budgets. Their biggest problem is the rent.

The total rent for their apartment is $900, which is split evenly among the three roommates. The rent is due on the last day of each month. The guys don't get paid until the first day of the next month.

Their landlord has no tolerance for late payments.

Each student had a different amount of money after being paid on April 1. At the end of that day, Cal had $1,100, Bernie had $800, and Doc had $600. As the month goes by, they each occasionally note how much they had left at the end of the day.

The table shows their records so far.

**Amount of Money
Remaining (in dollars)**

Date	Cal	Bernie	Doc
April 3	996	766	570
April 10	704	698	490
April 17	440	626	430

continued ▶

2. As of the morning of July 12, the Buck family has 100 pounds of coffee, while the Woods family has only 70 pounds. Each family consumes 5 pounds of coffee per day.

   a. On a single set of axes, make graphs showing how much coffee each family has remaining over the next ten days, beginning with the morning of July 12.

   b. How does the pair of graphs show that the Buck family starts with more coffee than the Woods family?

   c. How does the pair of graphs show that the two families are consuming coffee at the same rate?

   d. For each family, write an equation that describes the amount of coffee they have remaining after d days.

3. Is there a time when the two families are the same distance from the Green River? If so, when is it? How far are they from the Green River at that time? If not, why not?

4. Is there a time when the two families have the same amount of coffee? If so, when is it? How much coffee do they have at that time? If not, why not?

Graphing Calculator In-Outs

Some of the things you have been doing with paper graphs can also be done with technology.

You will probably decide that some problems are easier to do on a graphing calculator than on paper, whereas others are easier to do on paper. This activity will help you learn to use technology.

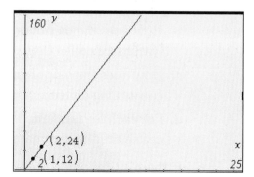

1. In *Previous Travelers,* you found a rule for estimating the number of pounds of beans needed for different numbers of people making the trip from Fort Laramie to Fort Hall. Here is a similar equation.

   Number of pounds of beans $= 12 \cdot$ (number of people)

   a. Enter and graph this function on a graphing calculator.

   Now use the trace feature to answer parts b and c.

   b. According to your graph, how many pounds of beans are needed for 20 people?

   c. A certain family brought 155 pounds of beans. According to your graph, how many people can they feed?

2. In *To Kearny by Equation,* you were given an equation for the profit that the Papan brothers made from their ferry service. That equation depended in part on the amount of the captain's travel time. Suppose the Papans decided to pay the captain $3 per day, regardless of the amount of business.

   In that case, the profit the brothers would get for one day could be determined by this equation.

   $$\text{profit} = 2W - 3$$

continued ▶

Here, W is the number of wagons that the ferry captain takes across the river.

a. Enter and graph this function on your graphing calculator.

Now use the trace feature to answer parts b and c.

b. How much profit do the Papans make when 25 wagons use their ferry?

c. How many wagons will the ferry have to carry for the Papans to make a profit of $25?

3. The In-Out table shown here is for the function
$$Y = 3X^2 - 7X + 2$$

Use technology to graph this function. Then use zoom and trace features to find the missing entries.

Where the *Out* is given, find all possible *In* values that will give the desired *Out*. If there aren't any, write "none."

Give your answers to the nearest tenth.

In	Out
1.31	?
−3.02	?
?	−1.04
?	−2.12
?	−2.05
8.57	?

Fort Hall Businesses

1. The Winstons were a large family seeking fortune in California, but they found the travels grueling. When they saw the beauty of Fort Hall and the opportunity to open a supply store in a rapidly growing community, they decided to settle there.

   From the time they opened the store, they had been able to add $50 to their bank account at the end of each month.

   They had some money in the account when they opened the store. Four months after the store opened, the account had grown to $360.

   a. As the family tried to clean up its bookkeeping procedures, they realized they had lost their original deposit slip. Based on the information above, determine how much money was in the account when they opened the store.

   b. Write a rule to express how much money would be in the Winstons' account, using x for the number of months since the store opened.

   c. Determine when the Winstons' account would reach $1,000. Explain your result.

2. There's a great show at the Grand Old Theater, and tickets are all the same price. Shortly after the box office opens, the manager is told that so far, the theater has sold 20 tickets, and the cash register now contains $40. A little later he is told that the theater has now sold a total of 60 tickets and the cash register now contains $70.

   a. How much money did the theater charge for each ticket?

   b. How much money was in the register before any tickets were sold?

   c. Write a rule for how much money was in the register after p people bought tickets.

   d. How much money was in the register after 250 people had bought tickets?

Sublette's Cutoff Revisited

In *Sublette's Cutoff*, you were given data showing the amount of water each of three families had at the end of certain days. You plotted the data and based your analysis on a pencil-and-paper graph.

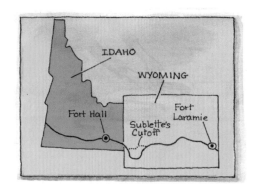

Now, you will reexamine that situation, with a slightly different goal and using a different technique.

Here are the data.

Gallons of Water Remaining

Family	Day 2	Day 5	Day 9
Jones	49	40	26
Sanford	68	48	25
Minto	30	21	14

For each family, use these steps to predict how much water they would have at the end of Day 15.

• Plot the data on a graphing calculator.

• Leave the data on the screen and graph a linear function that you think might approximate the data well.

• Examine how closely your function's graph approximates the data, and adjust the function until you think it approximates the data as well as possible.

• Use your final choice of function to make a prediction.

The Basic Student Budget Revisited

In *The Basic Student Budget,* you were given certain data about a situation and asked to make a prediction. You plotted the data and based your prediction on a paper graph.

Now you will reexamine *The Basic Student Budget* data using technology.

The savings and spending information about Cal, Bernie, and Doc—the roommates from *The Basic Student Budget*—is given below.

• The total rent for their apartment is $900, which is split evenly among the three roommates.

• The rent is due on the last day of each month. The guys don't get paid until the first day of the next month.

• At the end of payday April 1, Cal had $1,100, Bernie had $800, and Doc had $600.

• As the month goes by, they each occasionally note how much they have left at the end of the day. The table shows their records so far.
Here again is the technique you will use.

• Plot the data on a graphing calculator or other technology.

• Leave the data on the screen and graph a function that you think might approximate the data well.

Amount of Money Remaining (in dollars)

Date	Cal	Bernie	Doc
April 3	996	766	570
April 10	704	698	490
April 17	440	626	430

• Examine how closely your function's graph approximates the data. Adjust the function until you think it approximates the data as well as possible.

• Use your final choice of function to predict who will be able to pay rent on April 30.

On Your Own

When the Overland Trail travelers set out, they planned as well as they could, even though their information was limited.

In this POW, you will do some planning for a future stage of your own life—living on your own.

You may not yet have much information on what this part of your life might be like, so this is a research POW.

Imagine that you have just completed high school. There may have been adults who took care of many things for you before, but now you want to move out on your own.

For the purpose of this activity, assume that you need to provide your own financial support. What sorts of things do you need to plan for?

Be very detailed and accurate in your plan. If you are going to get your own apartment, then find an example of one and note the cost. You will need a job that you can enter with a high school education. You will need to know what you would get paid and how much of that is take-home pay.

It will probably be helpful to interview people for this POW. There is nothing like experience. What bills are you going to have to pay? Does your apartment rent include the cost of electricity? People already living on their own will be able to share with you how they manage the bills.

Your report should include a monthly budget, which is a plan showing how your money is going to be spent in a typical month.

Good luck!

continued ▶

○ *Write-up*

Because this is not a standard POW, you can't use the standard POW write-up. Use these categories instead.

1. *Description of the Task:* Explain in your own words what you are trying to do in this POW.

2. *Your Job:* You can consider these questions.
   - What is the job?
   - How do you find it?
   - What are your hours and salary?

3. *Your Living Arrangement:* Where would you live? Would you live by yourself or with roommates? What about furniture?

4. *A Monthly Budget:* Include more than just numbers. Discuss how and why you decided on your budget and where you got your information.

5. *Self-assessment:* What did you learn from this POW? In what ways do you think it will be helpful to you in the future?

All Four, One

In many activities, you have used different ways to represent a relationship between two variables.

- A situation
- A graph
- An In-Out table
- A rule for the table

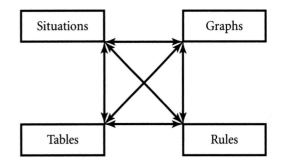

This gives four different ways to think about a single problem. The connections among these four forms of representation are some of the most fundamental ideas in mathematics.

Explain how the four representations—situations, graphs, In-Out tables, and rules—relate to one another. Use examples from this unit and examples of your own to show how you can go from one form of representation to another.

Travel on the Trail

As you leave Fort Hall, you travel toward the region northwest of Reno, Nevada. You hear of a man who runs an inn and trading post along the route. You plan to spend a few days visiting him. The innkeeper is James P. Beckwourth.

1. As of this morning, July 28, your traveling companions—the Barker family—are a few miles behind you. They have gone 23 miles beyond Fort Hall toward Beckwourth's trading post. Over the next portion of their journey, they will be traveling 12 miles a day.

   a. Make a graph that shows how far they will be from Fort Hall x days after July 28.

   b. Write a rule for your graph.

2. Your family is on its way from Fort Hall to Beckwourth's trading post. As of the morning of July 28, you are farther beyond Fort Hall than are the Barkers, but your family is moving more slowly. Hoping to arrive at Beckwourth's trading post before the Barkers, you must make some reasonable assumptions about the situation.

   a. Choose a specific value for your family's distance from Fort Hall as of the morning of July 28, and a specific value for your rate of travel. State the values you choose.

   b. Describe how the numbers you chose in part a will make the graph and rule for your family different from the graph and rule for the Barkers.

   c. Will the Barkers catch up to your family? If so, how might you find out when this would happen?

3. If Beckwourth's trading post is 140 miles away from where you are now, figure out the rate at which you will have to travel to arrive there 2 days before the Barkers. Use the distance you selected in Question 2a. *Note:* You may end up having to travel faster than the Barkers in order to do this.

About James Beckwourth

James Beckwourth, born around 1798 in Virginia, was the son of an African American woman and her master, a white man. At 19, during a fight with one of his slave bosses, he slugged his way to freedom and traveled from one end of the continent to the other.

On the trail, he fought with and against several Native American nations and served as a scout for the U.S. Army in the war against the Seminole. A contemporary of Kit Carson and Davy Crockett, Beckwourth was part of the brutal frontier tradition. He was said to have "fought and killed with ease and pleasure."

He also married often. After marrying a Crow woman, he was adopted into the Crow Nation and led the Crow in battle.

In 1850, Beckwourth located a pass through the Sierra Nevada into the American River Valley. This pass became a gateway to California during the gold rush. The mountain peak, the town, and the pass still bear his name.

Moving Along

Questions 1 through 3 use situations from this unit. The final problem uses the same mathematical ideas but without providing a situation.

1. In *Previous Travelers,* you worked with data showing the amount of beans needed (in pounds) by families of various sizes. Suppose your line of best fit goes through the points (4, 48) and (10, 120).

   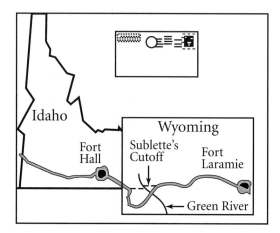

   a. Find an equation for this line.

   b. What is the rate of bean consumption per person? How is that value related to your equation? How did you find it from the data points?

2. In *Sublette's Cutoff,* you were given data showing the amount of water (in gallons) that the Minto family had left at the end of certain days. Suppose your line of best fit goes through the points (0, 36) and (6, 24).

   a. Find an equation for this line.

   b. What is the Minto family's rate of water consumption per day? How is that value related to your equation? How did you find it from the data points?

   c. How much water did the Minto family begin with? How is that value related to your equation? How did you find it from the data points?

continued ◗

3. *Who Will Make It?* provided data showing the distance the Fowler family was from the Green River in terms of the number of days they had been traveling. Suppose your line of best fit goes through the points (2, 300) and (10, 196).

    a. Find an equation for this line.

    b. What is the Fowler family's rate of progress per day? How is that value related to your equation? How did you find it from the data points?

    c. How far from the Green River was the Fowler family initially? How is that value related to your equation? How did you find it from the data points?

4. Relate the data below to a context.

    a. Create a situation for a problem with a graph that is a straight line that includes the points (3, 12) and (7, 32).

    b. Explain what the points (3, 12) and (7, 32) represent in your context.

    c. Find an equation for the line through these points,

    d. Explain how the numbers in the equation relate to your situation and how you found them from the data points.

All Four, One—Linear Functions

During this unit you have explored the connections among four representations for linear functions.

A **linear function** is a function with a graph that is a straight line. A common form for writing a linear function is $f(x) = ax + b$, where a is the **rate of change** and b is the **starting point.**

The four representations for relationships between two variables that have been explored in this unit are situations, tables, graphs, and rules.

Your Task

Create a how-to report for converting from any one of the four representations of a linear function to another. Begin by considering how many conversions you will have to address in your report. Be sure to discuss the role of a and b in each conversion.

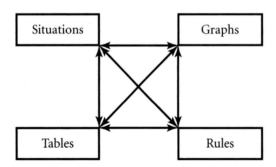

Straight-Line Reflections

In this activity, you will summarize key ideas about situations that lead to straight-line graphs. You will describe how the situation is related to the rule or graph.

1. What is it about a situation that makes its graph a straight line?

2. Choose a situation from either *Following Families on the Trail* or *Travel on the Trail*. Describe in words how you translated the situation into a graph.

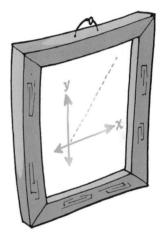

3. Graph the equation $y = 360 - 50(x - 4)$ on a graphing calculator or by hand. Notice this equation is linear.

   a. Use the tools you created in your how-to report to determine a linear equation in the form $y = ax + b$ from a graph or table.

   b. Demonstrate that the equation found in part a is equivalent to $y = 360 - 50(x - 4)$.

4. Graph the equation $3x + 2y = 9$ by calculator or by hand. Notice this equation is also linear.

   a. Use your tools to determine a linear equation in the form $y = ax + b$ from a graph or table.

   b. Demonstrate that the equation found in part a is equivalent to $3x + 2y = 9$.

Reaching the Unknown

As you continue your journey across the country, you will see how the use of variables, graphs, and equations helps solve challenges along the trail. Some of the new situations faced by the emigrants are complex—setting a schedule for watching the wagons at night based on who is available to stand guard or paying hired hands fairly based on their experience and the money available for salaries. Further variables complicate matters—the nights grow longer or shorter depending on the season, or families have a little more money to pay hired hands.

Some situations require you to find numeric values that fit more than one condition. Others require you to determine when two different situations will lead to the same result.

Once you've finally arrived in California, you'll see that life in this state wasn't all golden. You'll also see how the mathematics of expenses and profits played a role in the everyday decisions of early settlers.

Jeff Klein and Esteban Herevia map a route to California.

Fair Share on Chores

About 50 miles past Fort Hall, the California Trail splits from the Oregon Trail and heads into Nevada.

Two families, the Murphys and the Bensons, decided to continue on the Oregon Trail. You said good-bye and then ventured toward California.

Wagon trains often put their wagons in a circle to make a corral for the livestock. It was only in the movies that wagon trains created a circle to protect themselves from Native Americans.

Now that the Murphys and the Bensons have split from the wagon train, you have fewer wagons available.

The Washburn family decides that someone needs to keep an eye on their animals during the night. They announce that their children will take shifts each night, with one child at a time guarding the animals. Altogether, the animals need to be watched for ten hours. This family has two girls and three boys.

This sounds simple—two hours each. But the girls have other chores, and so do the boys. To balance out other assigned chores, the Washburn family decides that there should be one length of time for each girl's shift and another length of time for each boy's shift.

continued ▶

1. How long would you suggest that *each type* of shift be? Provide at least three different pairs of answers.

2. Using *G* to represent the length of each girl's shift and *B* to represent the length of each boy's shift, write an equation expressing the fact that the total of all their shifts is ten hours.

3. Suppose you know how long each girl's shift is. Describe *in words* how you could find the length of each boy's shift.

4. Write your sentence from Question 3 as a function, expressing *B* in terms of *G*. That is, write an equation that begins *B* = and has an expression using *G* to the right of the equal sign.

5. Graph the function from Question 4 on your calculator. Check to see if your answers from Question 1 are on the graph.

6. Use the trace feature on your calculator to find three more pairs of possible shift lengths from your graph.

Fair Share for Hired Hands

The Fulkerth family is large, and they have seven hired hands.

The family has a total of about $20 per week available for salaries.

Four of the hired hands are experienced at working on the trail. The other three are on their first trip.

It seems fair that the experienced hired hands should get more pay than those without experience. So the Fulkerths decide that there will be one rate for the four experienced workers and another rate for the three without experience.

1. What should each weekly pay rate be? Suggest three possible combinations. The salary total can be a few cents more or less than $20 if that helps you avoid fractions of a penny.

2. Plot your three combinations from Question 1. Use X for the pay rate of an inexperienced hired hand and Y for the pay rate of an experienced hired hand.

3. Connect the points with a straight line. Use this graph to find two more possible pairs.

4. Describe *in words* how you could compute the weekly pay rate for an experienced hired hand if you knew the rate for an inexperienced hired hand. Assume the weekly total is exactly $20.

5. Express your sentence from Question 4 as an equation, giving Y in terms of X.

6. Check to see if the two new pairs from Question 3 fit the equation from Question 5.

More Fair Share on Chores

As you saw in *Fair Share on Chores,* the Washburn family's two girls and three boys are responsible for watching the animals in shifts during the night.

After some experience, the family has decided that to balance out other chores, the shift for each boy should be half an hour longer than that for each girl.

1. They have realized that as the season gradually changes, the total amount of time needed for the shifts is not always ten hours. Therefore, they want to know about combinations of shift lengths with different totals.

   a. What are some possible combinations of shift lengths in which the shift for each boy is half an hour longer than that for each girl? Give four possibilities.

   b. Describe *in words* how you could find the length of each boy's shift if you knew the length of a girl's shift.

   c. Use your answer to part b to write an equation in which *G* represents the length of each girl's shift and *B* represents the length of each boy's shift.

   d. Graph your equation on the calculator.

   e. For each combination that you gave in part a, state how much *total time* will be covered by all the children combined.

2. On a particular evening, it turns out that ten hours of animal watching is required after all. Find a pair of shift lengths that would total ten hours and still have the shift for each boy be half an hour longer than the shift for each girl.

More Fair Share for Hired Hands

Once again, the Fulkerth family is planning its budget. Times are a little better now, but they haven't yet decided what the total budget should be for hired hands.

The same hired hands are still working for them. Now four hired hands are very experienced and three have only a little experience.

Although all the hired hands have some experience, the Fulkerths decide to continue having two pay rates. In the new pay scale, a very experienced hired hand will get $1 per week more than a less experienced hired hand.

1. Make several suggestions of how the Fulkerths could set up the two pay rates. Put these data in a table, with X representing a less experienced worker's weekly pay rate and Y representing a more experienced worker's weekly pay rate.

2. Graph the data from the table, and write an equation for your graph.

3. Find a set of pay rates that would make the total weekly pay for the hired hands approximately $30.

Water Conservation

Nevada seemed like a desert to the emigrants, who had been following large rivers most of the way from Westport. As you have seen, water was a very precious commodity on the Overland Trail. Travelers had to be careful not to run out.

They kept track of their water use, planning for the next opportunity to refill their water containers.

1. The Stevens family had a 50-gallon water container. In an effort to conserve water, they reduced their daily consumption to 3 gallons per day.

   If they began with a full container, how many gallons of water would they have left after 3 days? 8 days? 12 days? X days?

2. The Muster family was larger. They had a 100-gallon water container. Their daily consumption was 8 gallons per day. If they began with a full container, how many gallons of water would they have left after 3 days? 8 days? 12 days? X days?

3. Use your answers to the last part of Questions 1 and 2 to graph each family's water supply. Use *Number of days* for the horizontal axis and *Amount of water left* for the vertical axis. Graph both functions on the same set of axes.

4. Is there a time when both families would have the same amount of water left? If so, when would it happen, and how much water would both families have at that time?

5. In how many days would each family run out of water?

The Big Buy

Just like families along the Overland Trail, modern families need to plan for expenses. In this activity, parents pay their son and daughter to do chores.

Max and Jillian Verde want to make some money over their school's spring break. They ask their parents to let them work around the house to earn money.

Their parents agree, because Jillian and Max are saving to buy graphing calculators. Dad tells Jillian that he will give her a starting bonus of $10, and then pay her $5 an hour for the work she does around the house. Mom offers Max a slightly different deal. She will give him $40 to start, but only $3 an hour.

1. Write two separate equations—one for Jillian and one for Max—expressing how much money each will earn (including their starting money) in terms of time worked.

2. Graph both equations on the same set of axes.

3. If the graphing calculator costs $72, who will be able to buy a calculator with the least work time? Explain your answer.

4. If the graphing calculator costs $100, who will be able to buy one with the least work time? Explain your answer.

5. For what price must the calculator sell in order for Jillian and Max to earn that amount with the same number of hours of work? Explain your answer.

The California Experience

Who were the people in California in the 1850s?

Of course, Native Americans were there, probably for millennia. Some 300 different nations, including the Modoc, Washo, Maidu, Pomo, Cahuilla, and Miwok, held territory in what is now known as California.

Then came the Spanish. Out of their conquest and mixture with the Native Americans came a new culture and a new nation called Mexico, which became politically independent of Spain in 1821.

Indeed, that new Mexican nation claimed all of what is now the state of California, as well as all of Nevada and Utah and parts of Arizona, New Mexico, Colorado, and Wyoming. But in 1846, the United States provoked a war with Mexico in an attempt to gain territory. The United States won the war, which ended in 1848 with the Treaty of Guadalupe Hidalgo. In this treaty, Mexico was forced to cede much of what is now the southwestern United States.

Although the first wagon trains left Missouri for California in 1841, the great migration of the mid-nineteenth century was spurred by the discovery of gold in 1848, only nine days before the signing of the Treaty of Guadalupe Hidalgo.

continued ▶

The arrival of hundreds of thousands of gold seekers and others permanently changed the lives of the people who had been living in California. Settlements led to the destruction of whole nations of native peoples.

During the gold rush, thousands of people were brought from China to work as laborers. Although in 1850, there were only a few hundred Chinese in California, by 1852, about ten percent of the population was Chinese. Many of them lived in slavelike conditions.

Those who traveled on the California Trail in search of gold often ended up destitute, and a number of women resorted to prostitution to survive.

So the California experience was a mixture of many things. Only a very few became rich from the mining of gold.

Getting the Gold

Many of those who made the long trek to California were in search of gold. Although few were able to get rich, many tried.

One of the most common ways to find gold was to pan for it in streams.

To pan for gold, all a person needed was a $9 shovel, a $50 burro, and a $1 pan. A person could get an ounce of gold each day, on average, by panning.

One ingenious person discovered a way to get gold from a stream by using a trough. The trough was a long chute that miners set in the stream and rocked back and forth to separate the gold from the silt of the stream.

Although it was more expensive to get started with the trough method, that technique produced about twice as much gold each day as did the pan method. To use a trough, a person needed a team of two burros, a shovel, and a trough. The trough cost $311.

At that time, gold was worth $15 an ounce. The following questions involve the amount of profit (income minus expenses) from each method after a certain number of days. A loss of money is considered a negative profit.

1. How much profit will each method yield
    a. after 16 days?
    b. after 30 days?
    c. after 5 days?

continued ▶

2. Make two graphs on the same set of axes. One graph should show the profit from panning, and the other should show the profit from using a trough.

3. Find rules to show the profit using each method.

   a. Find a rule showing how much profit the panning method will yield after x days?

   b. Find a rule showing how much profit the trough method will yield after x days.

4. How many days will it take for a miner using each method to break even?

5. After how many days will the two methods yield the same amount of profit?

The Mystery Bags Game

Once a prospector accumulated some gold, he brought it to a government office for weighing. That office had a pan balance, like the one shown here, and a collection of lead blocks of known weights.

The officials would put the gold on one side and then try various combinations of the lead weights until the two sides balanced. For example, the result here shows that the bag of gold on the left weighs 23 ounces.

The Game

The officials came up with a game to pass the time when things got slow. One official would take one or more empty bags and fill them each with the same amount of gold. These bags of equal weight were called the *mystery bags*.

Next, the official would play around with ways to place the mystery bags and some lead weights on the pan balance so that the two sides balanced. The game was to figure out the weight of each mystery bag.

For example, this situation shows three mystery bags balanced with three different lead weights.

Your Task

If all the examples were like this one, the game would have been boring. The group made things more challenging by sometimes combining mystery bags and lead weights on either side of the balance.

continued ▶

See if you can solve these mystery bag puzzles by figuring out how much gold there is in each bag. Explain how you know you are correct. You may want to draw diagrams to show what's going on. You might consider how you could adjust what is on each side of the balance to simplify the situation but still be sure that the pans were equal in weight.

1. There are 3 mystery bags on one side of the balance and 51 ounces of lead weights on the other side.

2. There are 1 mystery bag and 42 ounces of weights on one side and 100 ounces of weights on the other side.

3. There are 8 mystery bags and 10 ounces of weights on one side and 90 ounces of weights on the other side.

4. There are 3 mystery bags and 29 ounces of weights on one side and 4 mystery bags on the other side.

5. There are 11 mystery bags and 65 ounces of weights on one side and 4 mystery bags and 100 ounces of weights on the other side.

6. There are 6 mystery bags and 13 ounces of weights on one side and 6 mystery bags and 14 ounces of weights on the other side. The playful government official could get in a lot of trouble for this one!

7. There are 15 mystery bags and 7 ounces of weights on both sides. At first, the official guessing thought this one was easy, but then he found it to be incredibly hard.

8. The official guessing the mystery bag weights wanted to be able to win easily every time, without calling you in for consultation. Therefore, your final task is to describe in words a procedure by which the official can find out how much is in a mystery bag in any situation.

More Mystery Bags

Here are some simple equations that might have come from mystery bag games. Find the weight of one mystery bag and explain how you got the answer.

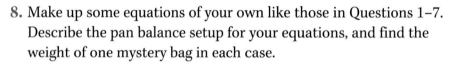

1. $M + 16 = 43$

2. $12M = 60$

3. $27 + 9M = 90$

4. $5M + 24 = 51 + 2M$

5. $43M + 37 = 56M + 24$

6. $12M + 13 = 5M + 62$

7. $5M + 2M + 100 = 15M + 20$

8. Make up some equations of your own like those in Questions 1–7. Describe the pan balance setup for your equations, and find the weight of one mystery bag in each case.

Here are some equations that you have developed in recent activities. In each case, try to adapt the methods you used above to find a value for x that makes the equation true.

9. $10 + 5x = 40 + 3x$

10. $50 - 3x = 100 - 8x$

11. $15x - 60 = 30x - 420$

Scrambling Equations

Usually, the concept of **equivalent equations** is used to make things simpler. But in this activity, you're going to make things more complicated. For example, look at the sequence of equations shown below.

$$x = 1$$

$$6x = 6$$

$$6x - 3 = 3$$

$$\frac{6x - 3}{2} = 1.5$$

All of these equations are equivalent, because they all have the same solution. You should be able to see what was done to each equation to get the one below it.

In this activity, you will begin by writing a *very simple* equation, such as $x = 1$. Then you'll write an equivalent equation that's more complicated, and then something equivalent to that, and so on.

This activity has some very precise rules. You will be changing your equation exactly three times. At each stage, you can do any one of these things.

- You can add the same integer to both sides of the equation.
- You can add the same multiple of x to both sides of the equation.
- You can subtract the same integer from both sides of the equation.
- You can subtract the same multiple of x from both sides of the equation.
- You can multiply both sides of the equation by the same nonzero integer.
- You can divide both sides of the equation by the same nonzero integer.

continued ▶

Remember that you are to do exactly three of these steps in any order. For instance, the example uses multiplication, then subtraction, and then division. At any time in the process, you're also allowed to do arithmetic steps to simplify the right side of the equation.

When you are done with this process, copy your final, complicated equation onto one side of a sheet of paper and put your original equation on the reverse side. This sheet will be exchanged with another student. You will then have the opportunity to "uncomplicate" someone else's scrambled equation.

More Scrambled Equations and Mystery Bags

Part I: More Scrambled Equations

This activity involves the same steps for getting equivalent equations that were described in *Scrambling Equations*.

1. The equations here show one sequence of three steps to "scramble" the equation $x = 3$.

$$x = 3$$
$$x - 5 = -2$$
$$10(x - 5) = -20$$
$$\frac{10(x - 5)}{4} = -5$$

   a. Describe what was done at each step.

   b. Check that $x = 3$ is a solution to the final equation in the sequence, and show your work.

For Questions 2 through 4, do two things.

   a. Uncomplicate each equation until you get back to a simple equation of the form "$x =$ some number."

   b. Take the value of x you get from the simple equation and substitute it back into the original equation to check that it makes the "complicated" equation true.

2. $3x - 5 = -2$

3. $\frac{x - 6}{4} + 1 = 7$

4. $4\left(\frac{x}{3} + 6\right) - 8 = 20$

continued ◗

Part II: More Mystery Bags

Earlier in this unit, you used the idea of a pan balance to solve mystery bag problems. The equations here might come from such problems. Solve them using the concept of equivalent equations, but also think about how each step you do is related to the pan-balance model.

5. $11t + 13 = 7t + 41$

6. $12 + 7w = 4w + 21$

7. $8(x + 3) + 19 = 15 + 2(x + 35)$

Family Comparisons by Algebra

In *Following Families on the Trail,* you were given some information about travel and coffee consumption for the Buck family and the Woods family. Here are the basic facts.

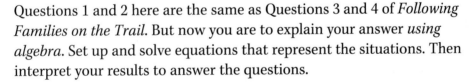

- As of the morning of July 12, the Buck family had gone 50 miles since leaving the Green River and was traveling at 15 miles per day.

- As of the morning of July 12, the Woods family had gone 10 miles since leaving the Green River and was traveling at 20 miles per day.

- As of the morning of July 12, the Buck family had 100 pounds of coffee and was consuming 5 pounds of coffee per day.

- As of the morning of July 12, the Woods family had 70 pounds of coffee and was consuming 5 pounds of coffee per day.

Questions 1 and 2 here are the same as Questions 3 and 4 of *Following Families on the Trail.* But now you are to explain your answer *using algebra.* Set up and solve equations that represent the situations. Then interpret your results to answer the questions.

1. Is there a time when the two families are the same distance from the Green River? If so, when is it, and how far are they from the Green River at that time? If not, explain why not.

2. Is there a time when the two families have the same amount of coffee? If so, when is it, and how much coffee do they have at that time? If not, explain why not.

Starting Over in California

Hundreds of thousands of people traveled to California in the middle of the nineteenth century.

Some came across the Pacific Ocean from China. Some sailed from the Atlantic coast to Panama, crossed land there, and then sailed again to California.

Still others, as you know, came by boat around Cape Horn or came by wagon or by horse on the Overland Trail.

Biddy Mason walked.

About Biddy Mason

Biddy Mason walked to California behind her master's 300-wagon train. Her job was to watch the cattle, but her master would not give her a horse, so she had to walk. She was one of the many enslaved African Americans brought to California by southern slave owners to work in the gold fields. Most remained enslaved.

Biddy Mason, however, broke away from her slave master and had the courage to sue for, and win, her freedom. She settled in California, working as a nurse and midwife. She became known for her generosity and great charity, taking in homeless children and supporting schools, churches, and hospitals.

Your Task

1. The Smith family had owned a blacksmith shop back in Pennsylvania. After arriving in California, the father, Richard, opened a blacksmith shop. The Smiths' main business was shoeing horses. They borrowed money from a friend to get the equipment and agreed to use all the income from horse shoeing to pay off this debt. After 4 weeks of payments, the Smiths owed $101. After 12 weeks, they had reduced the debt to $53. Assume the Smiths paid the same amount each week.

    a. How much did the Smiths borrow to open the shop?

    b. Find an equation that gives the amount they will owe after x weeks.

    c. How long did it take to pay off the debt?

continued ▶

2. Soon the Smiths heard about gold found in a river nearby. Richard decided to make pickaxes and shovels to sell to the miners. But Richard's daughter Kaley wanted to pan for gold. Richard made a deal with Kaley: "I'll let you pan for gold for one year. If after a year you make more money prospecting than I do selling pickaxes and shovels, I'll let you continue panning for gold. But if you make less than I do, you agree to return to the family business, stay at home, and work with us."

Surprise! On her first day of prospecting, Kaley found a gold nugget worth $150! After 2 months, Kaley had taken in $166 (including the $150), and her father had made only $42. After 5 months, Kaley's total earnings had reached $190, and her father's total was $105. If Kaley did not find any more large gold nuggets, do you think Kaley would get to keep panning for gold after a year?

3. The wagon train that your family is a member of has reached Sutter Creek in California. They decide to settle down in the area. Like Biddy Mason and the Smiths, you are now faced with the decision of how to make a living. Create a mathematical problem about some members of your family who either start a business or go for the gold.

Beginning Portfolios

California Reflections

The California Experience outlined some of the historical and social background of the gold rush.

Hien Nguyen writes a piece for her portfolio.

Write about your own feelings concerning this period of American history.

You may want to talk about the group or groups you identify with, about issues of justice or injustice, or about the process of social change.

You may also want to comment on how your ideas about the period have changed over the course of this unit.

Graphs Along the Way

The meaning and use of graphs played an important role in this unit.

Select one activity that illustrates how graphs can describe a problem situation.

Select another activity that illustrates how graphs can be used to make a decision about a problem situation.

Explain how each of these activities helped you understand the meaning and use of graphs.

The Overland Trail Portfolio

Now that *The Overland Trail* is completed, it is time to put together your portfolio for the unit.

- Write a cover letter that summarizes the unit.
- Choose papers to include from your work in the unit.
- Discuss your personal growth during the unit.

Cover Letter

Look back over *The Overland Trail* and describe the main mathematical ideas of the unit. This description should give an overview of how the key ideas were developed.

As part of compiling your portfolio, you will select activities that you think were important in developing the key ideas of this unit. Your cover letter should include an explanation of why you selected the particular items.

Selecting Papers

Your portfolio for *The Overland Trail* should contain these items.

- The items you selected in *Beginning Portfolios*

  Include your "California Reflection," to reflect the historical elements of the unit and your reaction to them. Also include the two activities about graphs that you selected, along with the explanation you wrote about how these activities helped you understand the meaning and use of graphs.

- *All Four, One—Linear Functions*

  This assignment is included because it summarizes the connections among situations, graphs, tables, and rules—four different ways of representing functions.

continued ▶

- An activity about starting value and rate

  The unit included problems about starting value and rate in several different contexts. Select an activity that illustrates these important points, and explain your selection.

- An activity about solving equations

  Several activities in this unit developed ways of thinking about and finding solutions to equations. Select one of these activities, and write about how it was meaningful for you.

- A Problem of the Week

  Select one of the three POWs you completed during this unit: *The Haybaler Problem, Around the Horn,* or *On Your Own.*

- Other quality work

  Select one or two other pieces of work that represent your best efforts. These can be any work from the unit—Problem of the Week, homework, classwork, presentation, and so forth.

Personal Growth

Your cover letter for *The Overland Trail* describes how the mathematical ideas developed in the unit. As part of your portfolio, write about your own development during this unit. You may want to address this prompt.

> Many of the problems in this unit involved graphs, and you learned how to use a graphing calculator to make graphs. Write about your reactions to using this tool.
>
> - What are the advantages of using the calculator to make graphs?
> - What are the advantages of doing graphs by hand?
> - How does each approach help your understanding of graphs and your ability to use them to solve problems?

Include any other thoughts you wish to share with a reader of your portfolio.

SUPPLEMENTAL ACTIVITIES

One of the supplemental activities for *The Overland Trail* extends a POW from this unit. Others continue your work with variables or build on the theme of the western migration. These are some examples.

- *More Bales of Hay* poses some questions related to *POW 6: The Haybaler Problem*.

- *Variables of Your Own* asks you to form meaningful expressions with variables.

- *Movin' West* poses a problem about the movement of people across the country.

- *Mystery Graph* is a graph of a nonlinear function. You are asked to find a number of different values for the function.

- *The Growth of Westville* presents situations that may seem to involve constant growth but that do not lead to linear graphs.

Pick Any Answer

Lai Yee has a new trick. He gives these instructions.

- Pick any number.
- Multiply by 2.
- Now add 8.
- Divide by 2.
- Subtract the number you started with.
- Your answer is 4.

1. Try Lai Yee's trick. Is the answer always 4? If you think it always is, explain why. If not, explain why it sometimes will be something else.

2. Make up a trick whose answer will always be 5.

3. Pretend that someone gives you a number that he or she wants to be the answer. Using the variable *A* to stand for that number, make up a trick whose answer will always be *A*.

Substitute, Substitute

For each of Questions 1 through 9, show both of these steps.

- Replace the variable by the value shown, writing the resulting expression in complete detail.

- Compute the numeric value of the expression you get in the replacement step.

Be sure to insert parentheses or multiplication signs where needed.

Note: The instructions in Questions 1 through 9 illustrate some of the many ways to describe the process of substitution. You should use the two steps of replacement and evaluation in each case.

1. Evaluate $5 + 6q$ at $q = 9$.

2. Find the value of $3z + 20$ when $z = -8$.

3. Get the numeric value of $15 - 4x$ for $x = -1$.

4. Evaluate $3t^2 + 7$ if $t = -2$.

5. What is $-r^2$ when $r = 8$?

6. Find $-w^2$ with $w = -6$.

7. Substitute $k = 3$ into $3 \cdot 2^k + 5$.

8. Evaluate $3a^3 + (4a)^2$ using $a = 5$.

9. Compare the results.

   a. Substitute $b = 6$ into $5b + 4 + 3b + 7$.

   b. Substitute $b = 6$ into $8b + 11$.

   c. Comment on what happened in parts a and b and explain why.

From Numbers to Symbols and Back Again

1. In *The Game of Pig,* you found the expected value in the one-and-one situation for a basketball player who makes 60 percent of her shots.

   You may have done similar problems using different percentages.

   Now consider the general case. Suppose the variable p represents the probability that a player succeeds with a given shot. Assume that this value is the same for every shot.

   a. Explain why the expected value for this player is $2p^2 + p(1 - p)$. Use the example of $p = 60\%$ as a model.

   b. Use the formula from part a to find the player's expected value for the case in which $p = .9$.

   c. What value of p would give an expected value of 1.5 points per one-and-one situation?

2. A quadrilateral has two diagonals. The diagram shows a five-sided polygon with five diagonals. It turns out, as you may already know, that a polygon with N sides has $\frac{N(N-3)}{2}$ diagonals.

   a. Based on this formula, how many diagonals does a 50-sided polygon have?

   b. How many sides would a polygon need to have so that it has at least 1000 diagonals?

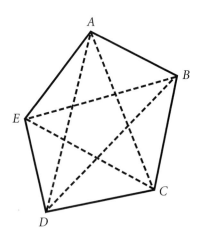

Classroom Expressions

Variables and summary phrases are useful in many situations other than those involving wagon trains.

In this activity, you will work with a set of variables that relate to a classroom setting. As in *Ox Expressions,* you should assume that the values of the variables are constant in all cases. For example, assume that every boy has the same number of pencils.

Reminder: Mathematicians often use **subscripts** so that they can use similar symbols to represent related quantities. You can think of a subscripted symbol as a two-letter variable for a single quantity.

For example, the two-letter symbols P_B and P_G are used in this activity to represent the number of pencils that each boy in the classroom has and the number of pencils that each girl in the classroom has. The subscript, either B or G, is written below and to the right of the main symbol, P. The combined symbol P_B is usually read *P sub B.* You need to exercise care in writing subscripted variables so that they don't look like the product of two separate variables.

Note: A symbol written *above* and to the right of the main symbol, which is like the way we write exponents, is called a **superscript.**

continued ▶

Symbol	Meaning
B	the number of **Boys** in the classroom
G	the number of **Girls** in the classroom
P_B	the number of **Pencils** each **Boy** has
P_G	the number of **Pencils** each **Girl** has
L	the cost of **Lunch** for each student (in cents)
S	the cost of a **Snack** for each student (in cents)
M	the amount of time each student spends in **Math** class per day (in minutes)
E	the amount of time each student spends in **English** class per day (in minutes)
H_M	the amount of time each student spends on **Homework** for **Math** per day (in minutes)
H_E	the amount of time each student spends on **Homework** for **English** per day (in minutes)

1. What, if anything, does each algebraic expression represent? Use a summary phrase, if possible.

   a. $B + G$

   b. GP_G

   c. $BM + BH_M$

   d. LS

2. Write an algebraic expression for each phrase.

   a. The total number of pencils for the students in the class

   b. The cost of lunch for the whole class

   c. The total amount of time students in the class spend on English each day, both in class and on homework

3. Make up some other meaningful expressions using this list of variables.

Variables of Your Own

1. Make up a set of between five and ten variables for a situation, the way you did in *Ox Expressions*.

   You might choose something like Marching Band Expressions, Baseball Game Expressions, Dating Expressions, or Clothing Store Expressions. Or you might prefer to make up a situation of your own.

   On the front of a sheet of paper, list your variables and what they stand for.

2. Below your list of variables, write three meaningful algebraic expressions using your variables.

   On the back of the same sheet of paper, write a summary phrase for each of your algebraic expressions.

3. On the front side of the same sheet of paper, write three summary phrases for which someone can write an algebraic expression using your variables. On the back side, write an algebraic expression for each of your summary phrases.

When you are ready, you will exchange papers with other students. Your task will then be to find summary phrases for each other's algebraic expressions and algebraic expressions for each other's summary phrases.

Painting the General Cube

Here is a problem you may have already worked on.

A cube with 5-inch edges is made up of smaller, 1-inch cubes.

Someone comes along and paints all the **faces** of the large cube, including the bottom. None of the paint leaks to the inside.

How many of the smaller cubes have only one face painted? How many have two faces painted? Answer the same question for three, four, five, and six faces.

Now answer the same questions for the situation in which the large cube is *N* inches long on each **edge.** The smaller cubes are again 1 inch on every edge.

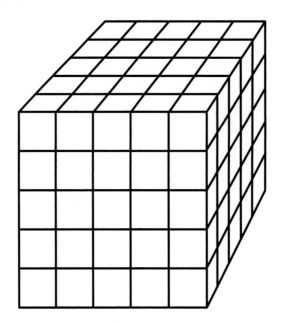

Integers Only

In *If I Could See This Thing*, you developed a formula for the number of people who would die on the trail between Fort Kearny and Fort Laramie.

You may have wondered why the formula could have 1.7 people dying. Part of a person does not make any sense.

There are many problems in which the function should only give whole-number outputs or only give integer outputs. Mathematicians have invented a special function and notation for dealing with such situations.

The function is called the *greatest integer function*. If the input to this function is represented by the letter N, the output is represented by the notation $[N]$.

This function is defined by "rounding down" the input, which means choosing the integer that is as large as possible without being larger than the input. If the input is already an integer, no rounding is needed. For example:

$$[7.2] = 7$$

$$[3] = 3$$

$$[-3.3] = -4 \text{ (Remember that } -3 \text{ is bigger than } -3.3.\text{)}$$

1. Which of these statements are true? If you think a statement is false, give a specific counterexample. If you think a statement is true, explain why.

   a. $[x + y] = [x] + [y]$

   b. $[x + 5] = [x] + 5$

   c. $[-x] = -[x]$

2. Draw the graph of the function defined by the rule *Out* = $[In]$.

More Bales of Hay

In *POW 6: The Haybaler Problem,* you were told that there were five bales of hay. They had been weighed two at a time in all possible combinations. You were given the weights for each of the ten possible pairs of bales. You found the weights of the five bales. That problem had a unique solution.

You might wonder whether there would have been a unique solution, or any solution at all, if the problem had used a different set of ten weights. That is, are there ten numbers that could not possibly stand for the weights of five bales of hay weighed two at a time in the different combinations?

Note: You may know that a typical bale of hay weighs about 100 pounds. In this problem, however, we want you to focus on the mathematical issues rather than on facts about hay. Therefore, you should think of bales of hay in the problem as representing arbitrary objects, whose weight could be much less or much more than that of a real bale of hay.

1. Before tackling the problem, consider this problem, which involves only three bales of hay.

    Three bales of hay were weighed two at a time in all three possible combinations. One combination weighed 12 pounds, one combination weighed 15 pounds, and one combination weighed 23 pounds. What were the weights of the three different bales? Does this problem have a solution? If so, is the solution unique? Explain your results and how you got them.

continued ▶

2. Now suppose that instead of 12 pounds, 15 pounds, and 23 pounds, the weights given in Question 1 had been different.

   a. Are there any sets of weights for the three pairs that would have made it impossible to solve Question 1? If so, for which sets of numbers is there a solution and for which sets is there no solution? Explain your answer.

   b. Which sets of weights for the three different pairs give *whole-number* solutions for the weights of the individual bales?

3. Now think about a five-bale problem, where there are ten pairs of bales to weigh. Could there be ten numbers that could not be the weights for the ten different pairs of the five bales? Justify your answer.

Spilling the Beans

Three travelers met one night along the Overland Trail. They decided to have dinner together.

Sam had seven cans of beans to contribute and Kara contributed five cans of beans. Jock didn't have any beans, but the three cooked up what they had. Each ate the same amount.

After dinner, Jock offered the 84 ¢ in his pocket and said that the other two could divide it up in an appropriate way. They all agreed that in this way everyone would have contributed a fair share to the dinner.

Jock thought that Kara's share of the money should be 35 ¢, but Sam and Kara convinced him that this was wrong.

1. Explain why Jock might have thought that Kara's share was 35 ¢.

2. Then explain what Kara's correct share should be.

From *Mathematics: Problem Solving Through Recreational Mathematics* by Averbach and Chein. Copyright 1980 by W.H. Freeman and Company. Adapted with permission.

More Graph Sketches

For each of the situations, sketch a graph that might represent what is happening. Include an appropriate scale on each axis.

1. The length of a burning candle as a function of the amount of time the candle has been burning

2. The weight of a person as a function of that person's age, over the course of a lifetime

3. The distance left to California as a function of the length of time since the wagon train left Westport

4. The height off the ground of a buffalo chip stuck to a wagon wheel as a function of time (Assume that the wagon is moving forward at a constant rate. Show the buffalo chip's height as the wheel makes three complete revolutions.)

5. Make up a situation of your own and sketch a graph.

Movin' West

The westward movement of people across the continent began well before the Overland Trail era.

In fact, the U.S. population has been moving westward since the earliest years of the country.

The "population center of gravity" of the United States is the point at which the country would balance if it were looked at as a flat plate with no weight of its own and every person on it had equal weight.

In 1790, this center of gravity was near Baltimore, Maryland. In 2000, it crossed the Mississippi River to Edgar Springs, Missouri (southwest of St. Louis).

1. From 1950 until today, the population center of gravity has moved about 50 miles west for every 10 years.

   Suppose this pattern continues for a while. Find a rule that expresses approximately how many miles west of Edgar Springs the population center of gravity would be when it is x years after 2000.

2. Edgar Springs is about 740 miles west of Baltimore. (It is also slightly south, but ignore that.)

   How does the rate of westward movement of the center of gravity between 1790 and 1950 compare with its rate from 1950 to 2000? Explain your answer carefully.

3. How long do you think the rule you found in Question 1 could continue to hold true? How do you think it might change?

From *Calculus* by Deborah Hughes-Hallett, Andrew M. Gleason, et al. (New York: John Wiley & Sons, 1994). Copyright 1994 by John Wiley & Sons, Inc. Adapted with permission.

What We Needed

Part I: Traveling Time

In the first part of this activity, you will figure out how long it took for your group's families to travel the entire distance from Fort Laramie to Fort Hall.

1. Based on your work in *Who Will Make It?*, you already know how many days it took the Fowler, Belshaw, and Clappe families to get from Fort Laramie to the Green River. Because the Belshaws arrived after the flood, add an additional 2 days to their travel time. It was typically an additional 18 days of travel from the Green River to the end of Sublette's Cutoff.

   If your family was mentioned in *Sublette's Cutoff,* you also know something about how long the journey from Fort Laramie to the end of Sublette's Cutoff took. The Jones and Minto families needed 26 days to reach the beginning of the cutoff, plus 15 days for the cutoff itself. The Sanfords used the same 26 days to reach the beginning, but needed 25 days for the cutoff, as their water resources grew scarce.

   For other families, use the fact that it took a typical wagon train on the main route 30 days to travel from Fort Laramie to the Green River and 18 days to travel from the Green River to the end of Sublette's Cutoff, for a total of 48 days.

   If your wagon train involved more than one family name, you should use the slowest time of the four families, because you would not leave them behind.

   Write down this time.

2. Find out how long it took your group's families to go from the end of Sublette's Cutoff to Fort Hall.

   Roll a pair of dice, find the sum, and add that to 8. This number will represent the average rate (in miles per day) at which your families traveled for this portion of the trip, which is a distance of 120 miles. You should have a rate between 10 miles per day and 20 miles per day.

continued ◗

Based on this rate, find out how many days it took from the end of Sublette's Cutoff to Fort Hall.

3. Add the results from Questions 1 and 2 to get the total number of days it took for your families to get from Fort Laramie to Fort Hall.

Part II: Supplies Needed

4. It turns out that each person in your group's families ate an average of 0.22 pound of beans per day between Fort Laramie and Fort Hall. Calculate the amount of beans each of your group's four families needed to bring on the trip.

5. It also turns out that each person used an average of 0.08 pound of sugar per day. Figure out how much sugar each of your group's families needed to bring.

Mystery Graph

The graph shows the variable y as a function of x, but it doesn't give a formula for this function.

Answer these questions based on the graph. State any assumptions you make about any portion of the graph that isn't visible. Give approximate answers if necessary.

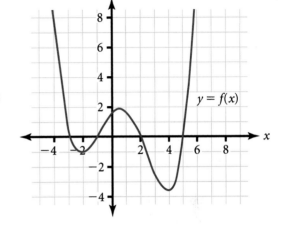

1. What number would you get for y if you used each of these values for x?

   a. $x = 4$

   b. $x = 0$

   c. $x = -1$

   d. $x = -4$

2. Find all the values of x for which y is 0.

3. For each of these values for y, find all the values for x that give that y-value.

   a. $y = 7$

   b. $y = 1$

   c. $y = -2$

   d. $y = -5$

4. Examine the part of the function between $x = -3$ and $x = 3$.

   a. Find the maximum point for the part of the function between $x = -3$ and $x = 3$. That is, what point with an x-coordinate between -3 and 3 has the largest y-coordinate?

   b. Find the minimum point for the part of the function between $x = -3$ and $x = 3$.

5. Find the values for x that give a positive value for y. Describe all possible answers.

High-Low Differences

Treat this activity like a Problem of the Week. You are to investigate a certain rule for generating a sequence of numbers.

The rule involves the repeated use of a three-step arithmetic process.

The following example shows how this three-step process works if you start with the number 473.

Step 1. Arrange the digits from largest to smallest: 743.

Step 2. Arrange the digits from smallest to largest: 347.

Step 3. Subtract the smaller number from the larger one.

$$
\begin{array}{r}
743 \\
- \ 347 \\
\hline
396
\end{array}
$$

The result of subtraction, 396, is called the **high-low difference** for the original number 473. The name comes from the fact that 743 is the highest number you can get from the digits of the number 473, and 347 is the lowest.

You can then take 396 and find *its* high-low difference, and then take that number and find *its* high-low difference, and so on. We will call the numbers you get in this manner the *high-low sequence* for the starting number 473.

Your task is to investigate these sequences for various starting numbers. You should continue with each high-low sequence until something interesting happens.

Begin by investigating three-digit starting numbers such as 473. Look for patterns in the high-low sequence and for reasons that explain what you see happening.

continued ▶

See what happens with four-digit numbers, five-digit numbers, and so on.

Your assignment has two components.

- Figure out as much as you can about high-low differences and high-low sequences.

- Explain as much of what you discover as you can.

Write-up

This POW is more like an exploration than a problem-solving process. Use the categories below for your write-up.

1. *Subject of Exploration:* Describe the subject you are investigating. What questions do you want to explore?

2. *Information Gathering:* Based on your notes, state what happened in the specific cases you examined. Include your notes with your write-up.

3. *Conclusions, Explanations, and Conjectures:* Describe any general observations you made or conclusions you reached. Wherever possible, explain why your particular conclusions are true. That is, try to *prove* your general statements. But also include *conjectures*, which are statements that you only *think* are true.

4. *Open Questions:* What questions do you have that you were not able to answer? What other investigations would you do if you had more time?

5. *Self-assessment*

High-Low Proofs

You may have come up with some interesting observations as you investigated *High-Low Differences.* But perhaps it's still a mystery to you why high-low differences work the way they do.

In this activity, you are to prove as many observations about high-low differences as you can.

Although you may want to use specific examples to illustrate your thinking, try to make your arguments as general as possible.

Keeping Track

1. The Perez family is keeping careful track of how much water they have at the end of each day. They had 120 gallons of water at the end of Day 5 and 100 gallons of water at the end of Day 10. They are using the same amount of water each day.

   a. How much water are they using per day?

   b. How much water did they start with?

   c. Write a rule for how much water they had after d days.

   d. Use your rule to find out how much water they had after 14 days.

2. Nina decided to save her money so she would be able to make the trip to California and start a new life. She opened a savings account and made an initial deposit. Then she added to her account, depositing the same amount of money at the end of each week.

   Nina's account had $50 in it after 10 weeks and $62 in it after 16 weeks.

   a. How much money did Nina deposit each week?

   b. How much money did she deposit in her savings account when she opened it?

   c. Write a rule for how much money she will have in the account after w weeks.

   d. Use your rule to find out how much money she will have in the account after 52 weeks.

A Special Show

The Grand Old Theater (from *Fort Hall Businesses*) is putting on a special show. Tickets for this show are more expensive than for the regular shows, but there is still only one ticket price.

As before, the cash register starts with some money in it to make change, and the manager checks the register occasionally. This time, he finds that after 30 people have bought tickets, the register has $50 in it, and that after 80 people have bought tickets, the register contains $110.

1. How much money did the theater charge per person?

2. How much money was in the register to start?

3. Write a rule for the amount of money that was in the register after *p* people bought tickets.

4. Use your rule to find out how much money the theater had after 250 people bought tickets.

5. If you haven't yet done so, graph the relationship between the amount of money in the register and the number of tickets that have been sold. How does your graph indicate both the amount of money in the register to start and the cost per person?

Keeping Track of Sugar

The Spencer family began their journey with a good supply of sugar. They checked their supply after 4 weeks on the trail and found that they had 80 pounds of sugar. They also noticed that after 14 weeks, they had 50 pounds of sugar. Assume that they consumed the same amount of sugar each week and didn't buy any additional sugar.

1. How much sugar did they use per week?

2. How much sugar did they start with?

3. Write a rule for how much sugar they had after *w* weeks.

4. Use your rule to find out how much sugar they had after 20 weeks.

5. If you haven't yet done so, graph the relationship between the amount of time that has gone by and the amount of sugar the Spencer family has left. How does your graph show how much sugar the Spencers started with and how much they used each week?

The Growth of Westville

As travelers moved across the country, some stayed in small towns along the way. Gradually, those towns grew larger.

Suppose Westville was established in 1850. The founders of this fictional town established its boundaries as a large square that was 2 miles on each side. As the population grew, the town council needed to extend the boundaries. They kept the town square, but increased each side by half a mile per year.

1. Make a table showing the area of Westville for each year from 1850 through 1860. Graph your data, using appropriate scales.

   Reminder: You can find the area of a square by squaring the length of a side. For example, in 1851, the side of the square was 2.5 miles long, so the area was $2.5^2 = 6.25$ square miles.

2. In 1850, Westville had 200 people. Each year for the next ten years, the population grew by 40 percent.

   Make a table showing the number of people in Westville for each year from 1850 through 1860. Graph your data, using appropriate scales.

3. In 1850, Westville had 200 people in a town that had an area of 4 square miles. There were 50 people per square mile. This is called the *population density*.

   From 1850 to 1860, the population grew. That by itself would make the density greater. But during the same period, the town's area also grew. That by itself would make the density smaller. You might wonder about the effect on density of these two factors combined.

   Make a table showing the population density for each year from 1850 through 1860. Graph your data, using appropriate scales.

Westville Formulas

The Growth of Westville describes the growth of a fictional town.

Your task in this activity is to create general formulas for the area, population, and density of Westville. Use N to represent the number of years since 1850. In other words, $N = 0$ for 1850; $N = 1$ for 1851; and so on.

Area

1. Recall that the boundaries of Westville formed a large square. In 1850, that square was 2 miles on each side. As the town grew, the founders of Westville decided to keep the town square, but they increased each side by half a mile per year.

   Based on this information, develop a general formula for the area of Westville in terms of N.

Population

2. In 1850, Westville had 200 people, and the population grew by 40 percent each year. Based on this information, develop a general formula for the population of Westville in terms of N.

continued ▶

Suggestion: Use the idea that each year's population is a fixed multiple of the population from the preceding year.

Density

3. Recall that *population density* means the number of people per square mile. Use your results from Questions 1 and 2 to develop a general formula for the population density of Westville in terms of *N*.

Big Picture

4. Compare these three formulas with the three graphs you created in *The Growth of Westville*. Describe how the general shape of each graph illustrates the mathematical relationships in the corresponding formula.

The Perils of Pauline

In 1869, the transcontinental railroad was completed. People could then travel westward by train, instead of by covered wagon. But trains could also be dangerous.

One day, for example, Pauline was walking through a train tunnel on her way to town. Suddenly, she heard the whistle of a train approaching from behind her!

Pauline knew that the train always traveled at an even 60 miles per hour. She also knew that she was exactly three-eighths of the way through the tunnel. She could tell from the train whistle how far the train was from the tunnel.

Pauline wasn't sure if she should run forward as fast as she could or run back to the near end of the tunnel.

She did some lightning-fast calculations, based on how fast she could run and the length of the tunnel. She figured out that whichever way she ran, she would just barely make it out of the tunnel before the train reached her. Whew!

How fast could Pauline run? Carefully explain how you found the answer.

The Pit and the Pendulum

Standard Deviation and Curve Fitting

The Pit and the Pendulum—Standard Deviation and Curve Fitting

Edgar Allan Poe—Master of Suspense

The title of this unit comes from a short story by Edgar Allan Poe (January 19, 1809–October 7, 1849). Poe was a poet, fiction writer, and literary critic. His contributions to literature include writing the first crime and detective stories and contributing to early science fiction.

Many of his stories involve mystery, suspense, and the bizarre, and "The Pit and the Pendulum" is no exception. It was published in 1842. Some of the folks on the Overland Trail probably read this story and others by Poe.

Lindsay Crawford and Catherine Bartz work on their group's initial experiments.

The Pit and the Pendulum

Excerpt from "The Pit and the Pendulum" by Edgar Allan Poe (1809–1849)

Looking upward, I surveyed the ceiling of my prison. It was some thirty or forty feet overhead, and constructed much as the side walls. In one of its panels a very singular figure riveted my whole attention. It was the painted picture of Time as he is commonly represented, save that, in lieu of a scythe, he held what, at a casual glance, I supposed to be the pictured image of a huge pendulum such as we see on antique clocks. There was something, however, in the appearance of this machine which caused me to regard it more attentively. While I gazed directly upward at it (for its position was immediately over my own) I fancied that I saw it in motion. In an  instant afterward the fancy was confirmed. Its sweep was brief, and of course slow. I watched it for some minutes, somewhat in fear, but more in wonder. Wearied at length with observing its dull movement, I turned my eyes upon the other objects in the cell . . . .

It might have been half an hour, perhaps even an hour (for I could take but imperfect note of time), before I again cast my eyes upward. What I then saw confounded and amazed me. The sweep of the pendulum had increased in extent by nearly a yard. As a natural consequence its velocity was also much greater. But what mainly disturbed me was the idea that it had perceptibly descended. I now observed—with what horror it is needless to say—that its nether extremity was formed of a crescent of glittering steel, about a foot in length from horn to horn; the horns upward, and the under edge evidently as keen as that of a razor. Like a razor also, it seemed massy and heavy, tapering from the edge into a solid and broad structure above. It was appended to a weighty rod of brass, and the whole hissed as it swung through the air . . . .

continued ▶

What boots it to tell of the long, long hours of horror more than mortal, during which I counted the rushing oscillations of the steel! Inch by inch—line by line—with a descent only appreciable at intervals that seemed ages—down and still down it came! . . .

The vibration of the pendulum was at right angles to my length. I saw that the crescent was designed to cross the region of the heart. It would fray the serge of my robe—it would return and repeat its operation—again—and again . . . .

Down—steadily down it crept . . . .

Down—certainly, relentlessly down! It vibrated within three inches of my bosom! . . .

I saw that some ten or twelve vibrations would bring the steel in actual contact with my robe, and with this observation there suddenly came over my spirit all the keen, collected calmness of despair. For the first time during many hours—or perhaps days—I thought. It now occurred to me, that the bandage, or surcingle, which enveloped me, was unique. I was tied by no separate cord. The first stroke of the razor-like crescent athwart any portion of the band, would so detach it that it might be unwound from my person by means of my left hand. But how fearful, in that case, the proximity of the steel! The result of the slightest struggle how deadly! Was it likely, moreover, that the minions of the torturer had not foreseen and provided for this possibility? Was it probable that the bandage crossed my bosom in the track of the pendulum? Dreading to find my faint, and, as it seemed, my last hope frustrated, I so far elevated my head as to obtain a distinct view of my breast. The surcingle enveloped my limbs and body close in all directions—save in the path of the destroying crescent.

Scarcely had I dropped my head back into its original position, when there flashed upon my mind what I cannot better describe than as the unformed half of that idea of deliverance to which I have previously alluded, and of which a moiety only floated indeterminately through my brain when I raised food to my burning lips. The whole thought was now present—feeble, scarcely sane, scarcely definite,—but still entire. I proceeded at once, with the nervous energy of despair, to attempt its execution.

For many hours the immediate vicinity of the low framework upon which I lay, had been literally swarming with rats. They were wild, bold, ravenous;

continued ▶

their red eyes glaring upon me as if they waited but for motionlessness on my part to make me their prey. "To what food," I thought, "have they been accustomed in the well?"

They had devoured, in spite of all my efforts to prevent them, all but a small remnant of the contents of the dish. I had fallen into an habitual see-saw, or wave of the hand about the platter; and, at length, the unconscious uniformity of the movement deprived it of effect. In their voracity the vermin frequently fastened their sharp fangs in my fingers. With the particles of the oily and spicy viand which now remained, I thoroughly rubbed the bandage wherever I could reach it; then, raising my hand from the floor, I lay breathlessly still.

At first the ravenous animals were startled and terrified at the change—at the cessation of movement. They shrank alarmedly back; many sought the well. But this was only for a moment. I had not counted in vain upon their voracity. Observing that I remained without motion, one or two of the boldest leaped upon the framework, and smelt at the surcingle. This seemed the signal for a general rush. Forth from the well they hurried in fresh troops. They clung to the wood—they overran it, and leaped in hundreds upon my person. The measured movement of the pendulum disturbed them not at all. Avoiding its strokes they busied themselves with the anointed bandage. They pressed—they swarmed upon me in ever accumulating heaps. They writhed upon my throat; their cold lips sought my own; I was half stifled by their thronging pressure; disgust, for which the world has no name, swelled my bosom, and chilled, with a heavy clamminess, my heart. Yet one minute, and I felt that the struggle would be over. Plainly I perceived the loosening of the bandage. I knew that in more than one place it must be already severed. With a more than human resolution I lay still.

Nor had I erred in my calculations—nor had I endured in vain. I at length felt that I was free. The surcingle hung in ribands from my body. But the stroke of the pendulum already pressed upon my bosom. It had divided the serge of the robe. It had cut through the linen beneath. Twice again it swung, and a sharp sense of pain shot through every nerve. But the moment of escape had arrived. At a wave of my hand my deliverers hurried tumultuously away. With a steady movement—cautious, sidelong, shrinking, and slow—I slid from the embrace of the bandage and beyond the reach of the scimitar. For the moment, at least, I was free.

The Question

Does the story's hero really have time to carry out his escape plan?

1. Based on the information you have, draw your own sketch of the prisoner's situation.

2. Go back through the story and carefully search for any additional information about the pendulum and the time for the prisoner's escape. Compile a group list of any information you find. If you are uncertain about the importance or relevance of a piece of information, write it down—you may need it later. Also write down any questions you have, and identify any information you wish you had.

3. In your group, share your initial opinions about the question.

Strict rules determine how knight pieces may move on a chessboard. Each "move" consists of two squares in one direction and then one square in a perpendicular direction. For example, knights may move forward (or backward) two squares and then to the right (or left) one square. Or they may move to the left (or right) two squares and then down (or up) one square.

Chess rules also state that no two chess pieces may occupy the same square at the same time, although knights may pass or jump over other pieces on the way to an empty square.

That's it—there are no other choices. So knights can get pretty bored spending their days on the chessboard.

One day two black knights and two white knights were sitting around on a 3-by-3 chessboard, just as you see here, feeling restless.

To liven things up, they decided to try to switch places. The white knights would end up where the black knights started out and vice versa. They could move only one at a time,

continued ▶

according to the rules of chess, and they had to stay within the nine squares of their board.

1. Can they do it?

2. If so, what is the least number of moves it will take them to switch? How do you know this number is the least?

3. If it is not possible, explain why not.

○ *Write-up*

1. *Problem Statement*

2. *Process:* Be sure to include a description of how you kept track of the various moves. Also describe your different approaches to working on the problem.

3. *Solution:* Be sure to explain why you think your answer represents the least possible number of moves or why you think the task is impossible.

4. *Extensions*

5. *Self-assessment*

Adapted from *aha! Insight,* by Martin Gardner, Scientific American, Inc./W.H. Freeman and Company, San Francisco © 1978. According to *The Penguin Book of Curious and Interesting Puzzles,* by David Wells (Penguin, 1992), this is one of the earliest known recreational chess problems; it was posed by Guarini di Forli in 1512.

Building a Pendulum

1. Tell the story of "The Pit and the Pendulum" to a family member, friend, or neighbor. Ask the listener if the time for the prisoner's escape plan seems realistic. Then ask how you could find out if it is.

2. Write about these topics.
   - What were the listener's reactions?
   - Did the listener think the amount of time in the story seemed realistic? What was the listener's reasoning?
   - How did the listener think you could find out how realistic this time estimate was?

3. Make a pendulum from materials that you find around your home.

4. Figure out a way to measure the period of your pendulum as accurately as you can. It may be helpful to work with the family member, friend, or neighbor. *Remember:* The **period** is the time it takes for your pendulum to swing back and forth once.

5. Describe how you measured the period.

6. Bring your pendulum to class.

Initial Experiments

You know that something determines the period of a pendulum, but it may not be clear exactly what that something is. Maybe there are several things.

In this activity, you will experiment to get an idea about what affects a pendulum's period.

1. Your teacher will assign you a variable from the list made in class. Do some experiments to see if that variable affects the period of a pendulum.

2. Prepare a written report, describing what you did, what observations you made, and what questions you still have.

Close to the Law

Zoe was doing a report on crime. Some people she interviewed believed that building more police stations would result in less crime. These people claimed that the closer you get to a police station, the less crime there is.

Others thought that nearness to a police station was not an important factor in a neighborhood's level of crime.

Zoe wanted to check out these competing claims. She called her local police station and got data on crimes in her area in the past year. She focused on robberies and how far each robbery had been from the station. She made this table.

Number of blocks from police station	Number of crimes per block
0–5	1.3
6–10	1.4
More than 10	1.6

1. Given this information, what relationship do you think there is between nearness to a police station and amount of crime? Explain your reasoning.

2. Zoe's brother Max thinks there might be factors affecting crime rate other than distance from the police station. List at least three other factors that might account for the differences in crime rates.

If I Could Do It Over Again

The process of experimentation often requires a person to test, refine the experiment, and then test again.

Even if an experiment does not yield the expected results, important knowledge often emerges from it, such as information about what not to do.

What advice would you give someone who is doing your group's pendulum experiment for the first time? Address these issues.

- Problems encountered in setting up the experiment
- Problems encountered in doing the experiment
- Materials you might have wanted to have
- Unexpected results

Time Is Relative

Nobody is a perfect timer. In this experiment, you will explore how accurately people can time things. Members of your group will take turns timing 5 seconds.

Here is how the experiment works.

- One person watches the second hand of the clock on the wall or on a group member's watch. A second person holds a stopwatch.

- The first person says, "Start," and then says, "Stop" after 5 seconds. The second person tries to start and stop the stopwatch on command so that it reads 5 seconds.

Naturally, you may be off a little bit each time you try it. Record your results to the nearest tenth of a second.

Take turns timing and recording the results.

What's Your Stride?

Your stride is the length of a typical step when you are walking normally at a steady rate. For this assignment, you should measure a stride from the front of one foot to the front of the other foot.

Give all measurements to the nearest inch.

1. Take a guess at the length of your stride, and write it down.

2. Measuring your own stride may not be easy.

   a. Think of a method for measuring the length of your own stride, and describe it clearly.

   b. Use your method to find the length of your stride.

3. Using your method from Question 2, find the length of a stride of each of two people not in your class. You might work with family members, neighbors, or friends. To avoid too wide a variation, you should work only with people who are teenagers or adults.

4. How do you think a **frequency bar graph** of the stride lengths of 50 people might look? Draw the graph based on your best guess.

5. Why might people want to know the lengths of their strides?

Pulse Gathering

If you repeatedly measure the same thing very carefully in the same way, will you get the same answer every time?

This activity provides a setting in which to look at this question.

1. Begin with your body at rest. Count the pulse beats at your wrist for a 15-second interval. Record your result as a whole number of pulse beats.

2. Repeat Step 1, again recording your result. Continue to repeat Step 1 until you have ten results. Some of these results may be identical.

3. In preparation for *Pulse Analysis,* share your data with everyone else in your group. Record each other's results. You should have ten results for each person in the group, including yourself.

Pulse Analysis

You should have a collection of data on the number of pulse beats in a 15-second interval—ten results for yourself and ten for each of your fellow group members.

1. Did you get the same result each time you counted your pulse beats for a 15-second interval? Why or why not?

2. a. Find the mean (average) of your own pulse data.

   b. Find the mean of the pulse data for your whole group.

   c. Was your group's mean the same as your own? Why or why not?

3. Draw frequency bar graphs that use individual bars for each measurement.

   a. Make a frequency bar graph of your own pulse data.

   b. Make a frequency bar graph for the pulse data of the whole group.

Corey Camel

Consider the case of Corey Camel—the enterprising, but eccentric, owner of a small banana grove in a remote desert oasis. Corey's harvest, which is worth its weight in gold, consists of 3000 bananas. The marketplace where the harvest can be sold is 1000 miles away.

Corey must walk to the market, but she can carry at most 1000 bananas at a time. Furthermore, being a camel, Corey eats one banana during each and every mile she walks (so Corey can never walk anywhere without bananas).

How many bananas can Corey get to the market?

○ *Write-up*

1. *Problem Statement*

2. *Process:* You will also work on a mini-POW that relates to this POW. In discussing your process on this POW, note how your work on *A Mini-POW About Mini-Camel* helped you. Be sure to discuss all the methods you tried in order to solve the POW itself.

3. *Solution*

   a. State your solution (or solutions) as clearly as you can.

   b. Do you think your solution is the best possible one? Explain.

   c. Explain how and why the answer to this POW is related to the answer to *A Mini-POW About Mini-Camel*.

4. *Extensions*

5. *Self-assessment*

Return to the Pit

This unit is complex and includes some investigations that are not directly concerned with the time the prisoner has to escape.

Reflecting on where you are will help you solve the unit problem.

Write answers to these questions for someone who knows nothing about Poe's story and who knows little about mathematics.

1. Write about the unit problem your class is trying to solve. State the goal as clearly as you can.

2. Write about what you have done so far and how that work will help you solve the unit problem. Be sure to explain clearly how **measurement variation** is involved.

3. Write down some questions you have about the unit and some points you don't yet clearly understand.

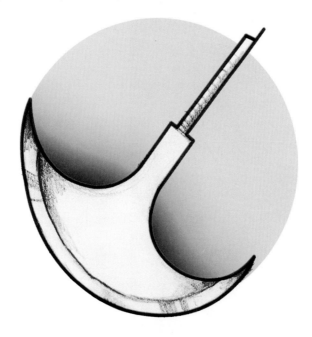

Statistics and the Pendulum

Now you have an idea of what this unit problem is all about. You probably have a list of variables that might affect the period of a pendulum. You've also seen that you can measure the exact same pendulum twice and get different periods. That makes things pretty unpredictable!

Scientists are used to uncertainty in their experiments. You might say that this uncertainty is "normal." In statistics, people have a very special meaning for the word *normal,* and they've come up with ways to describe how *abnormal* a particular measurement might be.

Lindsey Carvalho prepares a bar graph as an aid in analyzing the results from her experiment.

What's Normal?

You've seen some examples of normal curves. Now explore when the **normal distribution** might apply to real life.

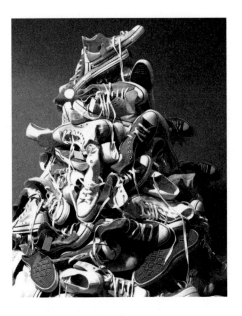

This activity describes several situations. You may not know what the real information is, so use your best judgment. You might make up a set of data that seems reasonable to you.

For each situation in Questions 1 through 4, follow these three steps.

a. Draw a frequency bar graph of the situation based on your idea of what the data might look like. Your graph should show axis labels and units of measurement. You will need to decide on scale intervals that are suitable to the situation.

b. Explain how you decided what the graph should look like. If you guessed, explain what made you guess the way you did.

c. State whether your graph appears to be approximately a normal distribution.

1. The number of people in your school who wear hightop sneakers, lowtop sneakers, dress shoes, or sandals to school on a given day

2. The frequency with which a 100-meter sprinter achieves certain times in running 200 races over the course of one year (Assume that the sprinter's average time is 12 seconds.)

3. The number of people in the United States who earn certain amounts of money (Use 300,000,000 as the total population of the United States. You might use categories such as "Income from $0 to $20,000," "Income from $20,000 to $40,000," "Income from $40,000 to $60,000," and so on.)

4. The number of people in the state of Hawaii who are of certain ages (Use 1,250,000 as the total population of Hawaii.)

A Mini-POW About Mini-Camel

Like Corey Camel, Mini-Camel owns a banana grove.

But Mini-Camel's harvest consists of only 45 bananas, and Mini-Camel can carry at most 15 bananas at a time.

The marketplace where Mini-Camel's harvest can be sold is only 15 miles away. Like Corey, Mini-Camel also eats one banana during each and every mile he walks.

1. How many bananas can Mini-Camel get to market?

2. Explain how Mini-Camel achieves this result.

3. Discuss how this problem is related to *POW 10: Corey Camel.* Explain how this mini-POW could help you solve POW 10.

Note: Your POW write-up asks you to refer to your work on this activity, so you'll need to keep a copy of your notes from this mini-POW.

Flip, Flip

Do the results of coin flips give a bell-shaped distribution? You can get an idea by performing some experiments.

1. Shake 10 coins together and let them fall. Record the number of heads. Do this experiment 15 times, recording the result each time, so that you get a total of 15 results. Each result is a number from 0 to 10.

2. Make a frequency bar graph showing the results of your experiments.

3. Do you think anyone in your class will have an experiment with a result of 0 (no heads)? With a result of 10 (all heads)? Explain your reasoning.

4. Predict what the class results will look like. Draw a frequency bar graph that you think will resemble the combined results from your class. Explain your reasoning.

5. Suppose you are given two coins and are told that one of them is unbalanced (but you don't know which one). You flip one of the coins 50 times, and it gives 28 heads and 22 tails. How confident would you be in deciding whether the coin you flipped is the unbalanced one? Explain your reasoning.

What's Rare?

Part I: Stride Lengths

Answer these questions, using the frequency bar graph from *What's Your Stride?*

1. Name a single stride length that you would categorize as ordinary. Name a single stride length that you would categorize as rare.

2. Where would you put the boundaries for each category? In other words, complete these sentences, replacing each "–?–" by a number.

   a. An ordinary stride length is from –?– to –?–.

   b. A rare stride length is less than –?– or greater than –?–.

3. Based on your answers to Question 2, estimate the answers to these questions.

   a. What percentage of all the data is in the ordinary category?

   b. What percentage of all the data is in the rare category?

Part II: Pulse Rates

Use the frequency bar graph from *Pulse Analysis* to answer these questions.

4. Categorize each measurement as ordinary or rare for the pulse rate of a person at rest.

   a. 20 beats for 15 seconds

   b. 17 beats for 15 seconds

   c. 12 beats for 15 seconds

   d. 28 beats for 15 seconds

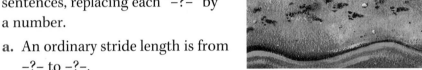

continued ▶

5. Where would you place the borderline for each category—ordinary and rare?

   a. An ordinary pulse rate is from –?– to –?–.

   b. A rare pulse rate is less than –?– or greater than –?–.

6. Based on your answers to Question 5, estimate the answers to these questions.

   a. What percentage of all the data is in the ordinary category?

   b. What percentage of all the data is in the rare category?

Part III: Timing of Five Seconds

Use the frequency bar graph from *Time Is Relative* to answer these questions.

7. Where would you place the borderline for each category—ordinary and rare?

   a. An ordinary result is from –?– to –?–.

   b. A rare result is less than –?– or greater than –?–.

8. Based on your answers to Question 7, estimate the answers to these questions.

   a. What percentage of all the data is in the ordinary category?

   b. What percentage of all the data is in the rare category?

Part IV: Comparing and Using Rarities

9. Compare the percentages you got in Question 8 with those you chose in Question 3 and Question 6.

10. Suppose you got a new stopwatch and used it to repeat the experiment of repeatedly timing 5 seconds. If you found that you had an average of 5.7 seconds after 10 timings, would you think the new stopwatch was defective? What if your average after 10 timings with the new stopwatch was 4.9 seconds? Explain your reasoning.

Penny Weight

Sarah and Tom's mom is a chemist, and one day she brought home a very sensitive scale. Sarah and Tom enjoyed learning how to use it.

One of the things they did was measure the weight of some pennies, one at a time. Here is a list of the results they got, arranged from lightest to heaviest. Weights are in milligrams.

2480	2484	2487	2490	2492	2495	2496
2497	2498	2499	2503	2503	2504	2505
2506	2507	2510	2511	2516	2517	

1. Given this information, what do you think is the best estimate for the weight of a penny? Why?

2. Sarah and Tom's Uncle Jack claimed that he had a counterfeit penny. Sarah and Tom didn't believe it was counterfeit, because it looked and felt real and because their uncle was always trying to fool them. They asked him if they could borrow the penny. When they weighed the coin, they got 2541 milligrams.

   Tom said the coin must be counterfeit because they never got a weight that high with their other pennies. Sarah isn't so sure it's counterfeit. She thinks that if they weighed it again, its weight might be closer to that of the weight of the others. Or she thought that if they measured more pennies, then Uncle Jack's coin might not seem so weird.

   What do you think, and why? If you don't think Uncle Jack's penny is counterfeit, then how heavy or light would a penny need to be before you believed it was counterfeit?

Mean School Data

Students at Kennedy and King High Schools were trying to determine what would affect the period of a pendulum.

At each school, students decided on standard pendulum characteristics for their initial experiments, including length, weight, and **amplitude.** Then, at each school, five groups took measurements of the period of the school's standard pendulum. Each group made the same number of measurements and then found the mean for those measurements. The tables give the mean pendulum periods found at each school.

Kennedy High

Group	Mean pendulum period (in seconds)
1	1.21
2	1.25
3	1.22
4	1.19
5	1.23

King High

Group	Mean pendulum period (in seconds)
1	1.16
2	1.22
3	1.31
4	1.11
5	1.30

continued ▶

1. Find the overall mean for each school's data.

2. One group from each school decided to test whether changing the amplitude at which the bob starts would change the period of a pendulum. They conducted the set of experiments again, but this time they used a different amplitude. Everything else stayed the same. Both groups now got a mean pendulum period of 1.29 seconds.

   If you were at Kennedy High, what would you conclude? If you were at King High, what would you conclude? In each case, explain your reasoning.

An (AB)Normal Rug

One day, Al and Betty got bored playing spinner games and decided to try rug and dart games.

Betty thought that playing on square or rectangular rugs would not be challenging enough, so she made some rugs that looked like normal distributions. As usual, each point in the rug had an equally likely chance of receiving a dart.

Each rug consisted of three parts—a central portion placed symmetrically between the other two. The three parts were separated by vertical lines that were equally distant from the center of the rug.

Al and Betty then tried to guess where a dart would land. One guessed it would land in the central portion of the rug. The other guessed it would land in one of the two outer parts.

1. These diagrams are like the rugs used by Al and Betty. The shaded area is in the central part of the rug, and the unshaded areas are the two outer parts. In each case, estimate what percentage of the area is shaded.

    a.

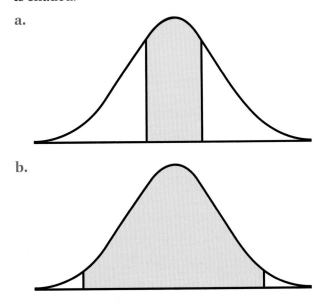

    b.

continued ▶

2. Trace each of these three rugs. Use two vertical lines to create a central region and two outer regions to fit the condition in the next paragraph. Shade the region in the center. Your shaded areas should be centered on each rug's vertical line of symmetry.

   Your task is to estimate where to put the vertical lines so that a player who guesses that the dart will land in the shaded area wins twice as often as a player who guesses it will land in the unshaded area. Explain how you determined that one area is approximately twice as large as the other.

   a.

   b.

   c.

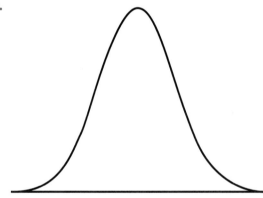

3. Make another copy of the three rugs in Question 2. Repeat the process described there, except this time make the player with the shaded region win 95% of the time. Again, explain how you estimated the areas.

Data Spread

Tai, Kai, and Mai understood how to find the mean of any set of data. They also knew that one set of data could have the same mean as another but look quite different.

On one occasion, they looked at these four sets of data, each of which has a mean of 20.

Set A	19, 19, 20, 20, 21, 21
Set B	10, 10, 20, 20, 30, 30
Set C	12, 13, 13, 27, 27, 28
Set D	9, 20, 20, 20, 20, 31

1. Based on your own intuition, arrange the four sets of data from the set that is "least spread out from the mean" to the set that is "most spread out from the mean." Explain your reasons for the order you choose. You may want to discuss this with others, but you should make your own decision.

Tai, Kai, and Mai were looking for a way to assign a numeric rating to measure how spread out from the mean a set of data was. They wanted a method in which the higher the rating, the more spread out the data set would be from its mean.

2. Tai liked the idea of using the **range** of the data to measure data spread. To find the range, you subtract the smallest number in the list from the largest one. For example, you find the range for set D by taking the difference $31 - 9$, which is 22.

   a. Find the range for each of the other sets of data.

   b. Based on Tai's method, arrange the four sets of data from the set that is least spread out from the mean to the set that is most spread out from the mean.

   c. Does your result from part b change your mind about your answer to Question 1? Explain.

You will learn about Kai and Mai's ideas in *Kai and Mai Spread Data.*

Kai and Mai Spread Data

In *Data Spread,* you saw Tai's idea for measuring data spread. His two friends had other suggestions. Here are the sets of data from that activity.

Set A   19, 19, 20, 20, 21, 21

Set B   10, 10, 20, 20, 30, 30

Set C   12, 13, 13, 27, 27, 28

Set D   9, 20, 20, 20, 20, 31

1. Kai proposed finding the distance of each number in the list from the mean and then adding those distances to get a measure for data spread. For example, in set C, because the mean is 20, the number 12 is 8 away from the mean. Similarly, each number 13 is 7 from the mean. So Kai would assign the numbers $8 + 7 + 7 + 7 + 7 + 8$, which is 44, to set C.

   a. Find the number that Kai would assign to each of the other sets of data.

   b. Based on Kai's method, arrange the four sets of data from the set that is least spread out from the mean to the set that is most spread out from the mean.

2. Mai's idea was to ignore the highest and lowest data items, removing just one item at each end, even if there were ties. Then she said that she would find the remaining data item that's farthest from the original mean and use that distance to measure data spread. For instance, with set B, Mai would drop the lowest number, one of the 10s, and the highest number, one of the 30s, leaving just 10, 20, 20, and 30.

continued ▶

Because the mean of set B is 20, the maximum distance from any of these numbers to the mean is 10. So Mai would assign the number 10 to set B.

a. Find the number that Mai would assign to each of the other sets of data.

b. Based on Mai's method, arrange the four sets of data from the set that is least spread out from the mean to the set that is most spread out from the mean.

3. Examine your answers to Questions 1b and 2b, as well as the answer to Question 2b of *Data Spread*. Whose measure of data spread—Tai's, Kai's, or Mai's—is closest to the intuitive guess you made in Question 1 of *Data Spread?* Explain your decision.

4. Invent a way to measure data spread that is different from these three. Describe how it works, and explain whether or not you think it is better.

The **standard deviation** of a set of data measures how "spread out" the data set is. In other words, it tells you whether all the data items bunch around close to the mean or if they are "all over the place."

The superimposed graphs show two normal distributions with the same mean. The taller graph is less spread out. Therefore, the data set represented by the taller graph has a smaller standard deviation.

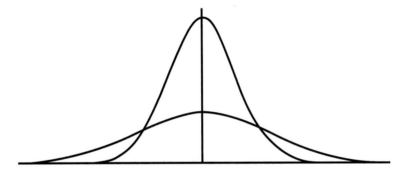

Calculation of Standard Deviation

Here are the steps for calculating standard deviation.

1. Find the mean.

2. Find the difference between each data item and the mean.

3. Square each difference.

4. Find the average (mean) of these squared differences.

5. Take the square root of this average.

Organizing the computation of standard deviation into a table can be very helpful. The table shows the calculation for these data: 5, 8, 10, 14, and 18. The mean is 11. The mean of a set of data is often represented by the symbol $\bar{x}$, which is read as "*x* bar."

continued ▶

The computation of the mean is shown to the left below the table. On the right below the table, step 4 is broken into two substeps—adding the squares of the differences and dividing by the number of data items.

The symbol usually used for standard deviation is the lowercase form of the Greek letter *sigma*, written σ.

x	$x - \bar{x}$	$(x - \bar{x})^2$
5	-6	36
8	-3	9
10	-1	1
14	3	9
18	7	49

sum of the data items = 55   sum of the squared differences = 104

number of data items = 5   mean of the squared differences = 20.8

$\bar{x}$ (mean of the data items) =
55 ÷ 5 = 11 σ (standard deviation) = $\sqrt{20.8} \approx 4.6$

Suppose you represent the mean as $\bar{x}$, use n for the number of data items, and represent the data items as x_1, x_2, and so on. The standard deviation can then be defined by the equation

$$\sigma = \sqrt{\frac{\sum_{i=1}^{n} (x_1 - \bar{x})^2}{n}}$$

The number inside the square root sign is the **variance** of the data set.

Standard Deviation and the Normal Distribution

The normal distribution was identified and studied initially by a French mathematician, Abraham de Moivre (1667–1754). He used the concept of normal distribution to make calculations for wealthy gamblers. That was how he supported himself while he worked as a mathematician.

The normal distribution applies to many situations besides those that are of interest to gamblers. (Measurement variation is one important example.) Therefore mathematicians have studied this distribution extensively.

continued ▶

When we use standard deviation to study the variation among measurements of a pendulum's period, we make this assumption:

Normality Assumption

If you make many measurements of the period of any given pendulum, the data will closely fit a normal distribution.

Some principles of standard deviation hold true for any normal distribution. Specifically, whenever a set of data is normally distributed, these statements hold true.

* Approximately 68% of all results are within one standard deviation of the mean.

* Approximately 95% of all results are within two standard deviations of the mean.

These facts can be explained in terms of area, using the diagram "The Normal Distribution."

The Normal Distribution

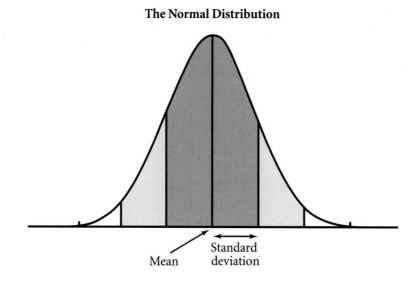

Mean

Standard deviation

The Normal Distribution

In this diagram, the darkly shaded area stretches from one standard deviation below the mean to one standard deviation above the mean. It is approximately 68% of the total area under the curve.

continued ◗

The light and dark shaded areas together stretch from two standard deviations below the mean to two standard deviations above the mean. They constitute approximately 95% of the total area under the curve.

Standard deviation provides a good rule of thumb for deciding whether something is "rare."

Note: In order to understand exactly where the specific numbers 68% and 95% come from, you would need to have a precise definition of normal distribution—a definition that is stated using concepts from calculus.

Geometric Interpretation of Standard Deviation

Geometrically, the standard deviation for a normal distribution turns out to be the horizontal distance from the mean to either of the two places on the curve where the curve changes from being concave down to concave up. In the diagram "Visualizing the Standard Deviation," the center section of the curve, near the mean, is concave down, and the two "tails" (that is, the portions farther from the mean) are concave up.

The two places where the curve changes its concavity, marked by the vertical lines, are exactly one standard deviation from the mean, measured horizontally.

Visualizing the Standard Deviation

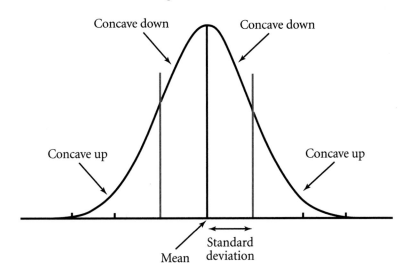

The Best Spread

Here are the four sets of data from *Data Spread.*

Set A   19, 19, 20, 20, 21, 21
Set B   10, 10, 20, 20, 30, 30
Set C   12, 13, 13, 27, 27, 28
Set D   9, 20, 20, 20, 20, 31

1. Write down the way you first intuitively arranged the four sets of data from the set that is least spread out from the mean to the set that is most spread out from the mean.

2. Calculate the standard deviation of each set of data. Arrange the four sets of data from the set with the smallest standard deviation to the set with the largest standard deviation.

3. Are the two arrangements from Questions 1 and 2 the same? Explain how they are different and why they might be different.

4. Recall that Tai thought that *range* was a good method for measuring data spread. Make up two new sets of data, set X and set Y, in which set X has a larger standard deviation than set Y but set Y has a larger range than set X.

Making Friends with Standard Deviation

You will be working with the concept of standard deviation to decide which variables actually have an effect on the period of a pendulum. It will be helpful for you to become familiar with what standard deviation means.

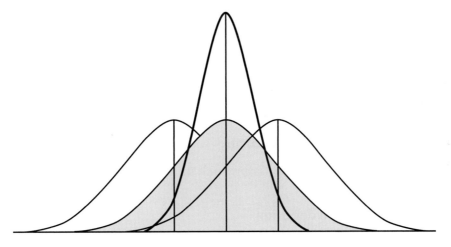

1. First explore what happens to the mean and the standard deviation of a set of data when you add the same number to each member in the set.

   a. As a group, make up a set of five numbers that are all different. Find the mean and the standard deviation of your set.

   b. Now choose a nonzero number and add it to each member of your set. Find the mean and the standard deviation of your new set.

   c. Repeat part b, using a different nonzero number. Add this number to each member of your original set of data, and find the mean and standard deviation of the new set. Keep repeating this process until you see patterns, and then describe those patterns.

   d. Explain why your pattern should occur.

      ○ Explain why the mean changes as it does when you add the same thing to each member of the set.

      ○ Explain why the standard deviation changes as it does when you add the same thing to each member of the set.

continued ▶

2. Now explore what happens to the mean and the standard deviation of a set of data when you multiply each member in the set by the same number.

   a. Begin with the same set of data as in Question 1a. Choose a nonzero number other than 1. Multiply each member of your set by that number and find the mean and the standard deviation of the new set.

   b. Choose another nonzero number other than 1, and repeat what you did in part a.

   c. Keep choosing new nonzero numbers to use as multipliers for each member in your set. Find the mean and the standard deviation of each new set until you see patterns. Describe those patterns.

   d. Explain why your patterns occur.

3. Make up a set of data that satisfies each of the given conditions as closely as you can.

   a. Mean, 6; standard deviation, 1

   b. Mean, 10; standard deviation, 1

   c. Mean, 7; standard deviation, 2

Deviations

1. Find the mean and the standard deviation of these data.

    24, 25, 15, 19, 17

Your task in the rest of this activity is to make up new data sets, each having either the same mean or the same standard deviation as the data in Question 1.

If you can, do these problems without actually calculating the mean or the standard deviation of each new data set. Explain how you know without calculating that the data set fits the conditions.

2. Make up a set of five data items that has the same mean as the data in Question 1 but that has a smaller standard deviation.

3. Make up a set of five data items that has the same mean as the data in Question 1 but that has a larger standard deviation.

4. Make up a set of five data items that has the same standard deviation as the data in Question 1 but that has a different mean.

Eight Bags of Gold

Once upon a time there was a very economical king who gathered up all the gold in his land and put it into eight bags. He made sure that each bag weighed exactly the same amount.

The king then chose the eight people in his country that he trusted the most and gave a bag of gold to each of them to keep safe for him. On special occasions he asked them to bring the bags back so he could look at them. He liked looking at his gold even though he didn't like spending it.

One day the king heard from a foreign trader that someone from the king's country had given the trader some gold in exchange for some merchandise. The trader couldn't describe the person who had given her the gold, but she knew it was someone from the king's country. Because the king owned all the gold in his country, it was obvious that one of the eight people he trusted was cheating him.

The only scale in the country was a pan balance. This scale couldn't tell how much something weighed, but it could compare two things and indicate which was heavier and which was lighter. The person whose bag was lighter than the others would clearly be the cheat. So the king asked the eight trusted people to bring their bags of gold to him.

Being very economical, the king wanted to use the pan balance as few times as possible. He thought he might have to use it three times to be sure which bag was lighter than the rest. His court mathematician thought that it could be done in fewer weighings. What do you think?

To answer this question, follow these steps.

1. Develop a scheme for comparing bags that will always find the light one.

2. Explain how you can be sure that your scheme will always work.

continued ▶

3. Explain how you know that there is no scheme with fewer weighings that will work.

Note: Each comparison counts as a new weighing, even if some of the bags are the same as on the previous comparison.

○ *Write-up*

1. *Problem Statement*

2. *Process:* Describe how you found your answer and how you convinced yourself that your method works in all situations. If you think your answer is the best possible, describe how you came to that conclusion.

3. *Solution:* Describe your solution to the king's problem as clearly as possible. Write a proof that your method will work in every situation. If you think that the king cannot find the lighter bag in fewer than three weighings, prove it.

4. *Extensions*

5. *Self-assessment*

Penny Weight Revisited

In *Penny Weight,* you saw that Sarah and Tom had been weighing a bunch of pennies on a sensitive scale. Here again are their results (in milligrams).

2480	2484	2487	2490	2492	2495	2496
2497	2498	2499	2503	2503	2504	2505
2506	2507	2510	2511	2516	2517	

1. Compute the mean and standard deviation of these weights. Record all your computations clearly so you can compare results with others in your group.

    Remember the steps in finding the standard deviation.

    a. Find the mean.

    b. Find the difference between each data item and the mean.

    c. Square each difference.

    d. Find the average (mean) of these squared differences.

    e. Take the square root of this average.

2. Reconsider the problem of the penny that Sarah and Tom's Uncle Jack claimed was counterfeit. When Sarah and Tom weighed that penny, they got a weight of 2541 milligrams.

    a. Based on your results in Question 1, what can you say about the probability that a real penny would have a weight this far from the mean?

    b. Do you think that Uncle Jack's penny is real or counterfeit?

Can Your Calculator Pass This Soft Drink Test?

1. A soft drink company sells its beverage in 1-liter bottles. One liter is equal to 1000 milliliters. The abbreviation for milliliter is mL.

   The machine that fills the bottles is not perfect. The amount of soft drink it puts into the bottles fits a normal distribution, with a mean of 1000 mL and a standard deviation of 5 mL.

   If the machine puts more than 1005 mL into a bottle, the bottle may spill when opened, causing customers to complain. If the machine puts less than 995 mL into a bottle, the amount in the bottle will appear less than it should be, causing customers to feel cheated.

   A quality-control worker checks the filled bottles before they are sealed to see if they fit within these bounds. If a bottle is too full or not full enough, the worker removes the bottle from the assembly line.

   Approximately what percentage of bottles get removed from the assembly line?

continued ▶

2. A manufacturer of graphing calculators keeps track of the length of time it takes before the product is returned for repair. She finds that the mean is 985 days and the standard deviation is 83 days.

   She wants to set a time period during which her company will warranty the calculators—that is, a period in which the company will replace the calculators at no cost to the customer if the products do not function properly. She does not want to have to replace more than 2.5% of those sold.

   Assume that the number of days before calculators need repair is normally distributed. How many days' warranty would you advise her to give her customers? Explain your reasoning.

3. Students' scores on a college entrance examination follow a normal distribution, with a mean of 490 and a standard deviation of 120. The college of your choice, Big State University, accepts only students whose scores on this test are 600 or higher.

   Estimate the percentage of students who are eligible for admission to Big State on the basis of their test scores, and explain your reasoning.

A Standard Pendulum

It's been a while since you actually measured the period of a pendulum. You now have some statistical tools. Normal distribution and standard deviation will help you understand the variation in measurements that can occur, even if you don't change the pendulum.

Over the next few days, you will get back to measuring. You will use these tools to decide "what matters." In other words, you will figure out, statistically, what variable or variables really affect the period of a pendulum.

Isaiah Henderson creates a misleading graph by changing the scale on the x-axis.

The Standard Pendulum

The pendulum that is illustrated here will be called the *standard pendulum* for the rest of the unit.

Here are the specifications for this standard pendulum.

- Weight of bob   1 washer
- Length of string   2 feet
- Amplitude   20°

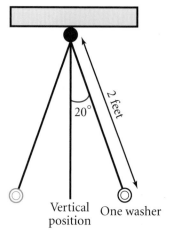

As you investigate what determines a pendulum's period, you will be looking at pendulums that differ in some respect from this one. You will compare their periods to the period of this one.

Find the period of this standard pendulum using the procedure agreed upon by your class, and record your result.

Repeat the experiment, again recording the period. Continue gathering more data as time allows.

Standard Pendulum Data and Decisions

You should have a large collection of measurements made for the period of the standard pendulum.

Although all groups worked with the same specifications for the pendulum, their results for the period probably were not all alike. In fact, each group probably came up with slightly different values each time they measured the period, even though the pendulum itself didn't change.

1. Why do different experiments using the exact same pendulum give different values for the period?

2. Make a frequency bar graph of the data you have. Choose intervals that you think are appropriate for grouping the data.

3. Draw a normal curve that approximates the graph you made in Question 2. Your curve should go approximately through the tops of the bars of your frequency bar graph.

4. Based on either your frequency bar graph from Question 2 or the curve from Question 3 (or both), estimate the mean and the standard deviation for the data.

5. Suppose you built a pendulum using different specifications, measured its period using the same procedure as in *The Standard Pendulum,* and got a result different from the mean you found in Question 4. How far from the mean would you need that new period to be before you were confident that the difference was not simply due to measurement variation? Explain your answer.

Pendulum Variations

You have seen that you may get slightly different results each time you measure the period of the standard pendulum.

As noted in *Standard Deviation Basics,* you are making an assumption about these measurement variations.

Normality Assumption

> If you make many measurements of the period of any given pendulum, the data will closely fit a normal distribution.

In *Standard Pendulum Data and Decisions,* you estimated both the mean and the standard deviation of this normal distribution.

In this activity, you will look at what happens to the period if the pendulum is changed in certain ways. You will then use the results from this activity to decide what factor or factors seem to determine the pendulum's period.

In conducting your experiments, you should measure the period in exactly the same way as you did in *The Standard Pendulum.* For example, if in that activity you measured the time for 10 swings and then divided by 10 to get the period, you should do the same thing here.

Make each measurement twice, but treat each as a separate result. Do not average your two measurements.

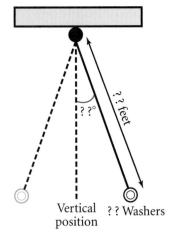

? ?°

? ? feet

Vertical position

? ? Washers

continued ▶

In each experiment, your pendulum should be the same as the standard pendulum except for the characteristic being studied. Here again are the specifications of the standard pendulum.

- Weight of bob   1 washer
- Length of string   2 feet
- Amplitude   20°

1. Changing the Weight

Measure the period of a pendulum that is the same as the standard pendulum except that it has a weight of 5 washers.

2. Changing the Length

Measure the period of a pendulum that is the same as the standard pendulum except that it is 4 feet long.

3. Changing the Amplitude

Measure the period of a pendulum that is the same as the standard pendulum except that it has an amplitude of 30°.

A Picture Is Worth a Thousand Words

Some say a picture is worth a thousand words. However, pictures, like words, can sometimes be misleading. This can be the case when people use graphs to make a point.

Graphs are pictures that convey information. Just as there are people who forget to read the fine print in text, there are also people who don't look carefully at graphs to see what the numbers are really telling them. These people can be tricked into reaching false conclusions.

Here is an example. A television station wants to convince its advertisers that viewers are changing over to that channel in huge numbers. The station has experienced slow but fairly steady growth over the past year. If the station made a simple month-by-month graph of its ratings, this is what the graph might look like.

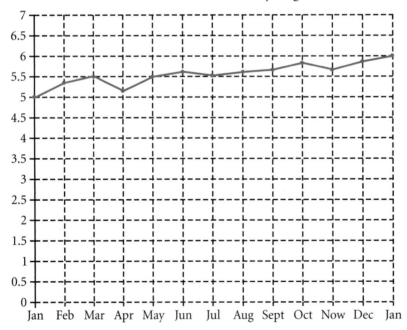

The Math Channel Makes Steady Progress!

continued ▸

As you can see, the station has increased its viewers by 1 ratings point, from 5.0 last January to 6.0 this January. But there are many ways the graph can be changed to give the impression that the station is doing even better than it actually is. One way is to show only part of the graph. The version below depicts only the changing part of the graph. It gives the appearance of a larger increase than does the first graph because the reader cannot see what the change relates to. The reader sees the graph starting at the bottom of the picture and ending at the top.

The Math Channel Adds Many Viewers!

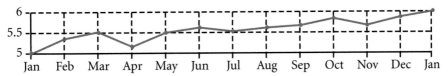

The graph can be made even more dramatic by changing the scales of the axes so that the distance between 5.0 and 6.0 is greater. For instance, if you change the vertical scale so that each interval is worth 0.1 instead of 0.5, here is what the graph will look like.

The Math Channel Is Hot!

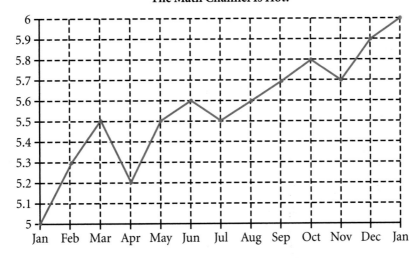

continued ▶

Changing the horizontal axis can help too. In the next graph, the change is reported only every 3 months, so there are no "downs," only "ups." Making the graph line bolder also adds to the effect.

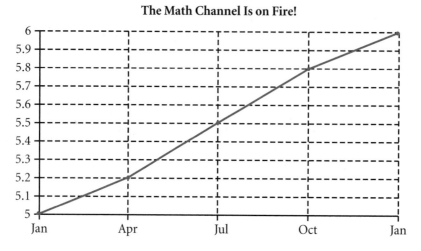

The Math Channel Is on Fire!

Your Assignment

1. Create a misleading graph of your own. This requires that you make two graphs: an original graph and a misleading graph

2. Describe the impact you were trying to create in Question 1. Explain how your graph is misleading.

3. Cut out a graph from a newspaper or magazine and write about why you think it is fair or why you think it may be misleading.

Pendulum Conclusions

1. In *Pendulum Variations,* you looked at what happened to the period of a pendulum when you changed from the standard pendulum. Based on those experiments, you have probably reached some conclusions about what factor or factors determine the period of a pendulum.

   Summarize your conclusions clearly. Support your conclusions using the concepts of normal distribution and standard deviation.

2. In the experiments from both *The Standard Pendulum* and *Pendulum Variations,* you had to work closely with your fellow group members.

   Decide what grade each member of your group deserves for the work he or she did in the past few days. Be sure to grade yourself.

   In assigning a grade to each person, you may want to consider these factors.

   - The suggestions the person made
   - How well the person listened to others
   - How supportive the person was to others
   - Whether the person helped the group stay on task
   - Whether the person helped the group when it got stuck
   - Whether the person helped settle disagreements in the group
   - Whether the person helped the group reach a consensus

POW Revision

Tonight you will revise your write-up of *POW 11: Eight Bags of Gold,* based on the feedback given to you by your classmates.

Your revision should be on paper *separate from your original.* It can take many forms, depending on the feedback you received.

* You can redo your entire write-up.
* You can rewrite certain sections that need refinement.
* You can include additional comments and diagrams that would improve or clarify parts of your write-up.

After revising your write-up, write a paragraph on the value of reading other students' papers and getting feedback on your own paper.

Bring these four items to the next class.

* Your original write-up
* The reviews you received from your classmates
* Your revisions
* Your evaluation of the experience of reading other students' write-ups and receiving feedback on your own write-up

Twelve Bags of Gold

Here we are, back with our economical king. Thanks to your class's work on *POW 11: Eight Bags of Gold,* he found the thief in a very economical manner. Because he has been so economical, he now has even more gold.

The 8 bags got too heavy to carry, so he had to switch to 12 bags. Of course, each of his 12 bags holds exactly the same amount of gold as each of the others, and they all weigh the same. Well, . . . maybe not.

Rumor has it that one of his 12 trusted caretakers is not so trustworthy. Someone, it is rumored, is making counterfeit gold. The king sent his assistants to find the counterfeiter. When they found her, she wouldn't tell them who had the counterfeit gold she had made, no matter how persuasive they were.

All the assistants learned was that one of the 12 bags had counterfeit gold and that this bag's weight was different from the others. They could not find out from her whether the different bag was heavier or lighter.

The king needed to know two things.

- Which bag weighed a different amount from the rest?
- Was that bag heavier or lighter?

Of course, he wanted the answer found economically. He still had only the old pan balance scale. He wanted the solution in two weighings, but his court mathematician said it would take three weighings. No one else could see how it could be done in so few weighings. Can you figure it out?

continued ▶

Find a way to determine which bag is counterfeit and whether it weighs more or less than the others. Use the pan balance as few times as possible. Keep in mind that what you do after the first weighing may depend on what happens in that weighing. For example, if the scale balances on the first weighing, you might choose bags for the second weighing different from the bags you would choose if the scale does not balance on the first weighing.

○ *Write-up*

1. *Problem Statement*

2. *Process:* Based on your notes, describe what you did.
   - How did you get started?
   - What approaches did you try?
   - Where did you get stuck?
   - What drawings did you use?

3. *Solution:* Because this problem is much more difficult than *POW 11: Eight Bags of Gold,* you shouldn't be too disappointed if you don't get a solution using just three weighings. Your task is to give the best solution that you found and to explain fully how your solution works.

4. *Extensions*

5. *Self-assessment*

Graphs and Equations

You're closing in on the unit problem. Now you know *what* determines the period of a pendulum, but you still need to figure out the relationship between the period and the controlling variable.

Pretty soon you'll gather some more data and look for a formula. In preparation for that, you're going to do a graphing "free-for-all"—an open-ended investigation of equations and their graphs. That exploration will give you an idea of the kind of formula to try.

Irvin Vasquez, Nery Guzman, Vincent Torres, and Jason Badua use their calculators to explore the graphs of a variety of functions.

Maliana the Market Analyst

Maliana is a market analyst. She studies the relationship among an item's price, the number of items sold, and the amount of profit.

A high price does not necessarily mean high profit. If a company makes a great product but charges too much for it, the company will not sell very many units and therefore won't make very much profit.

On the other hand, a low price does not necessarily mean making high profit either. If the company sells the product for too low a price, it will sell a lot of units but will not be making much profit.

You can see the complexity. The ideal price is somewhere in the middle—not too low and not too high. Maliana studies the market and finds the most profitable price at which to sell an item.

You have hired Maliana to analyze the market and suggest prices to charge in your music store. You have asked her to provide information about the price to charge for compact discs and compact disc players. Maliana comes back with this rather incomplete information.

continued ▸

Compact Discs		Compact Disc Players	
Price charged (in dollars)	Expected profit (in dollars per month)	Price charged (in dollars)	Expected profit (in dollars per month)
8	900	20	−600 (loss of $600)
10	2000	30	900
18	2200	40	1500
20	1600	70	1350
22	400	80	300

For each item (compact discs and compact disc players), do these things.

1. Sketch a graph showing the relationship between different prices and the expected profit.

2. Sketch a line or curve that you believe represents the relationship between price and profit.

3. What price do you think will maximize the profit for each item? Explain your reasoning.

Birdhouses

Mia and her classmates in carpentry class had spent the semester constructing birdhouses. Today Mia was in charge of painting. Her group had painted two of the birdhouses after 1 hour of work, and six of them after 3 hours.

1. How many birdhouses do you think they will have painted at the end of 8 hours?

2. How can you generalize your answer to Question 1?

You probably made some very natural assumptions when you answered Question 1. It's reasonable to assume that the group painted the same number of birdhouses every hour, especially because the data you have so far fit that assumption.

But what if you received additional information that contradicted that assumption?

3. Suppose you found out that at the end of 5 hours, Mia and her group had actually painted only nine birdhouses. Now what would you predict for the total for 8 hours? Explain your reasoning using a graph, table, or equation.

So Little Data, So Many Rules

1. Consider an In-Out table with just one pair.

$$In = 2, Out = 5$$

   a. Find three rules that fit this pair. At least one of your rules should be nonlinear.

   b. For each rule you find, find three pairs of numbers that fit that rule.

   c. Graph each rule that you made up.

2. Repeat the steps in Question 1 for an In-Out table that has just this pair.

$$In = 4, Out = 2$$

Graphing Free-for-All

In this activity you will use a graphing calculator or other technology to explore the graphs of a variety of functions and equations. The understanding of graphs and their equations that you gain will help you find the period of the 30-foot pendulum.

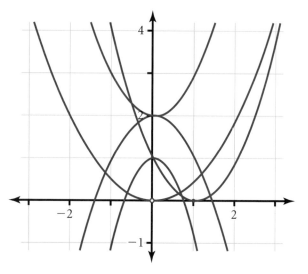

Take careful notes on the graphs you examine. You will be learning about other functions during your classmates' presentations on this activity, so you will also need to take notes on their presentations. In *Graphing Summary,* you will summarize your own conclusions as well as information from other presentations.

For each equation you look at, your notes should include these four kinds of information.

- The equation
- A sketch of the graph
- The viewing window you used on the graphing calculator or the scale you used with other technology
- An In-Out table, found by tracing the graph and by substituting into the equation

You may find it helpful to put each example on a separate sheet of paper.

Graphs in Search of Equations I

The coordinate plane shows three graphs, labeled *a*, *b*, and *c*. For each graph, do these things.

- Find the coordinates for four points that lie on that graph.
- Put the four pairs of coordinates into an In-Out table.
- Write an equation for that table.
- Check that your equation seems to work for all the points on the graph.

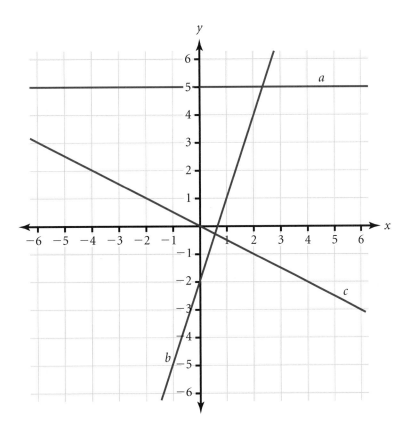

Graphs in Search of Equations II

As in *Graphs in Search of Equations I,* this coordinate plane shows three graphs. For each graph *a* through *c,* complete these steps.

- Find the coordinates for four points that lie on that graph.
- Put the four pairs of coordinates into an In-Out table.
- Write an equation for that table.
- Check that your equation seems to work for all the points on the graph.

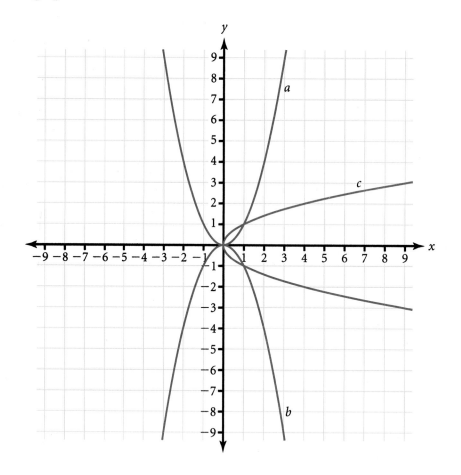

Graphing Summary

Through your work on *Graphing Free-for-All,* you should now have a collection of information about different equations and their graphs. Your task now is to organize and summarize this information.

Create a summary document that will help you find an equation that fits a particular graph, or sketch the graph for a particular equation.

Measuring and Predicting

Almost there! You're about to gather some data about the periods for pendulums of different lengths. This will set the stage for the final task—analyzing the data and making a prediction for the 30-foot pendulum.

You might want to think about how you will test your prediction.

Daniel Barajas, Eric Baraias, and Nery Guzman find equations to match their parabolas.

An Important Function

Based on earlier work, you have determined that the period of a pendulum seems to be a function of its length.

You will now gather some data about that function. Use the standard weight (1 washer) and standard amplitude (20°), but vary the length.

For each length that you examine, find the time for 12 periods, because the prisoner in Poe's story thought there were about 12 swings remaining when he created his plan to escape.

Graphs in Search of Equations III

For each of the graphs *a* and *b*, complete these steps.

- Find the coordinates for four points that lie on that graph.
- Put the four pairs of coordinates into an In-Out table.
- Write an equation for that table.
- Check that your equation seems to work for all the points on the graph.

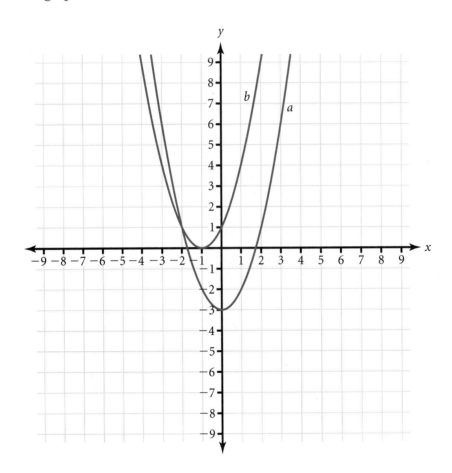

The Thirty-Foot Prediction

Now that you have the data on the time required for 12 swings for pendulums of several different lengths, your task is to make a prediction for a 30-foot pendulum.

Look for a function f that fits all of your data as well as possible. You probably won't find a function that fits the data perfectly, but do the best you can.

Once you are satisfied with your choice of function, find $f(30)$. That is, find out what your function would predict as the time required for 12 swings of a 30-foot pendulum.

Now work with your classmates to build Poe's pendulum and test your prediction.

Mathematics and Science

Think about what you have learned in this unit, both in mathematics and in science.

Choose what you consider to be the two or three most important ideas or concepts.

For each idea or concept, complete these steps.

- Write what you learned about that idea or concept.

- Write about a problem for which it might be useful to know that idea or concept.

Beginning Portfolios

1. State the central problem for this unit in your own words. Then write a careful description of how you arrived at a solution. This description will serve as the cover letter for your portfolio.

2. In *Mathematics and Science,* you identified two or three key ideas or concepts from this unit.

   a. Describe the role that those ideas or concepts played in solving the unit problem.

   b. Select a class or individual activity that helped you understand each idea or concept. Describe what you learned in those activities.

Carla Garcia prepares work for her portfolio.

The Pit and the Pendulum Portfolio

Now that *The Pit and the Pendulum* is completed, it is time to put together your portfolio.

Cover Letter

Your work on Question 1 of *Beginning Portfolios* will serve as your cover letter.

Selecting Papers

Your *The Pit and the Pendulum* portfolio should contain these items.

• Activities selected in *Beginning Portfolios*

Include your written work for Question 2 as well as the activities you selected.

• *Mathematics and Science*

• *Graphing Summary*

• A Problem of the Week

Select one of the four POWs you completed during this unit (*The Big Knight Switch, Corey Camel, Eight Bags of Gold*, or *Twelve Bags of Gold*).

• Other quality work

Select one or two other pieces of work that represent your best efforts. These can be any work from the unit—Problem of the Week, individual or group activity, or presentation.

continued ▸

Personal Growth

In addition to the papers you selected, discuss these issues.

- How you liked doing science experiments in a mathematics class
- What ideas you learned about in the unit but would like to understand better
- How you felt about grading yourself and others on group work
- Any other thoughts you might like to share with a reader of your portfolio

SUPPLEMENTAL ACTIVITIES

Many of the activities in *The Pit and the Pendulum* involve experiments and data gathering. In analyzing data from these experiments, you used the concepts of normal distribution and standard deviation. Another important theme of the unit is the study of functions and their graphs. Experiments, data analysis, and functions are some of the themes in the supplemental problems as well. These are some examples.

- In *Height and Weight,* you are asked to plan and carry out an exploration of the relationship between two variables.

- *Making Better Friends* and *Mean Standard Dice* strengthen your understanding of mean and standard deviation.

- In *Family of Curves,* you look at how changes in a function lead to changes in its graph.

Poe and "The Pit and the Pendulum"

Learn more about Edgar Allan Poe, the author of the short story "The Pit and the Pendulum." What was his life like? What else did he write?

Read the entire story to discover how the prisoner came to find himself strapped to the table and what happened to him after he freed himself from the danger of the descending pendulum. You may recall that the excerpt closes with the words "For the moment, at least, I was free."

Report on what you learned about Poe and the story.

Getting in Synch

Al and Betty are taking a break from their probability games and have gone to the circus to ride on the Ferris wheels. There are several different Ferris wheels at the circus. Al chooses one of them, and Betty chooses another.

Al and Betty each get on at the bottom of the Ferris wheels' cycles, and the two Ferris wheels start at the same time.

For Al's Ferris wheel the period is 40 seconds—that is, it takes 40 seconds for it to make a complete turn. Betty's Ferris wheel has a period of 30 seconds.

1. How long will it be until the next time Al and Betty are together at the bottoms of their Ferris wheels?

2. Redo Question 1, using each of the following combinations of periods for the two Ferris wheels.

   a. Al's: 40 seconds; Betty's: 25 seconds

   b. Al's: 31 seconds; Betty's: 25 seconds

   c. Al's: 23 seconds; Betty's: 18.4 seconds

3. Is it possible to find periods for the two Ferris wheels for which Al and Betty will never be at the bottom at the same time again?

4. What generalizations can you make?

Height and Weight

In class, you are exploring how the period of a pendulum may be affected when different variables, such as length and weight, are changed.

In this activity, you will deal with variables in a different context. Your task is to choose a setting in which to examine the relationship between *weight* and *height*. Design and carry out a plan for exploring how weight might depend on height in that context.

1. State the context of your exploration. For example, you may want to study how the weight of an animal depends on its height.

2. Describe a plan for this exploration.

   a. How will you gather information?

   b. What other variables will you take into account?

   c. What methods will you use to analyze your data?

3. Carry out your plan. Gather and study your data and come to some conclusions.

4. Describe your conclusions, evaluate the strengths and weaknesses of your plan, and discuss the difficulties involved in developing a connection between height and weight.

Octane Variation

Elizabeth and her dad decided to find out if the type of gasoline they used affected their car's mileage. They decided to try three different grades of gasoline and to measure the number of miles per gallon their car got for each.

When they began the experiment, the tank was filled with 87-octane gasoline. Elizabeth's dad drove until the tank was nearly empty. When he filled up the tank, he wrote down the octane of the gasoline he'd just been using, the number of gallons he'd used since the last fill-up, and the number of miles he had driven since the last fill-up.

He then went through the same procedure for each of the other two types of gasoline. Here is the information he presented to Elizabeth.

Octane used	Number of gallons used	Number of miles traveled
87	9.5	304
89	8.6	292
92	9.2	322

1. Given this information, do you think that the higher-octane gasoline yields better mileage? Explain your reasoning.

2. Elizabeth's brother Zeke thought there might be some other explanations for the variation in miles per gallon besides the octane of the gasoline. List all the other possibilities you can think of.

3. In what ways could Elizabeth and her dad have carried out a better experiment?

Stem-and-Leaf Plots

Frequency bar graphs provide one way to picture numeric data. A
stem-and-leaf plot is similar to a frequency bar graph. These plots use
a special system of grouping data and provide additional information
within each group.

A Sample Plot

To illustrate, consider this data set of 20 numbers.

4.7, 6.8, 6.2, 5.4, 3.4, 3.7, 4.9, 5.2, 4.0, 5.4,

4.3, 7.2, 4.5, 4.8, 2.9, 5.5, 4.8, 5.1, 5.8, 4.2

These numbers might be the heights, in centimeters, of a group of
plants, each measured after a specific length of time.

Because these numbers are given to the nearest tenth, the *stem* for each
number is its whole-number part, and the *leaf* for each number is its
decimal part. If the first few items were put into a frequency bar graph,
grouped by whole-number part, the result might look like this.

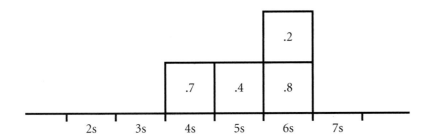

Here, the labels within the individual boxes indicate what data items
they represent. The .7 box above 4s represents the data item 4.7;
the .8 box above 6s represents the data item 6.8; and so on.

continued ▶

A stem-and-leaf plot presents the same information like this.

Stem	Leaf
2	
3	
4	7
5	4
6	8, 2
7	

Here is a complete stem-and-leaf plot for the 20-item data set. The leaves are arranged in increasing size, rather than left in the order in which they appeared in the original list.

Stem	Leaf
2	9
3	4, 7
4	0, 2, 3, 5, 7, 8, 8, 9
5	1, 2, 4, 4, 5, 8
6	2, 8
7	2

The length of the list for each stem provides the same type of information that is in a frequency bar graph—that is, the number of data items in a given group. This plot shows, for example, that the majority of the data items were "4 point something" or "5 point something." The stem-and-leaf plot has the advantage over a frequency bar graph that it gives details of the data items within each group.

Choosing Stems and Leaves

In the data set just illustrated, all numbers were given to the nearest tenth, so the whole-number part became the stem and the decimal part became the leaf for each number. For data sets with a different numerical range, a different way may be needed to separate numbers into stems and leaves. For example, consider this 15-item set (which might represent student scores on a test):

83, 54, 71, 95, 86, 92, 83, 65, 81, 73, 87, 90, 63, 85, 78

continued ◗

In this set, the tens digit of each number would be its stem and the ones digit would be its leaf. For instance, the first item, 83, would have 8 as its stem and 3 as its leaf. Because the scores range from the 50s to the 90s, the stems would be the numbers 5, 6, 7, 8, and 9. The stem-and-leaf plot would look like this, with the leaves left in their original order.

Stem	Leaf
5	4
6	5, 3
7	1, 3, 8
8	3, 6, 3, 1, 7, 5
9	5, 2, 0

Note: Context is needed to clarify the numeric values. For instance, just from looking at this stem-and-leaf plot, you would not know whether the first entry represented 54 or 5.4.

Comparing Two Data Sets

For two data sets from a similar setting, a back-to-back stem-and-leaf plot can provide a comparison. For instance, suppose this was a second set of test scores.

86, 91, 75, 83, 77, 95, 84, 74, 96, 84, 77, 82, 85, 63, 92

The results from the two data sets can be compared by placing one set of leaves to the left of the stems and the other set to the right of the stems. The result could look like this, with the leaves arranged in numeric order outward from the stems.

Second data set Leaf	Stem	First data set Leaf
	5	4
3	6	3, 5
7, 7, 5, 4	7	1, 3, 8
6, 5, 4, 4, 3, 2	8	1, 3, 3, 5, 6, 7
6, 5, 2, 1	9	0, 2, 5

The leaves on the left are somewhat farther down compared with the leaves on the right. This suggests that students did better on the second test than on the first.

Quartiles and Box Plots

The median of a set of data splits the set into two equal portions, with half the data at or above the median and half the data at or below the median. This idea can be extended using the concept of *quartiles* to create four sections of equal size within the data set. These four sections can be represented visually using a *box plot.*

Quartiles

Just as the median splits a data set in half, each half can be split in half again, creating four equal-sized data sets. To illustrate, consider this 25-item data set, in which the numbers have been arranged by increasing size.

22, 23, 23, 23, 24, 24, 25, 25, 25, 26, 27, 27, 28,

29, 30, 30, 31, 31, 32, 32, 32, 32, 33, 35, 39

These numbers might be measurements of stride length, in inches. Because this data set has an odd number of items, the median is the middle item, 28. There are 12 data items below 28 and 12 data items above 28.

Each of these 12-item sets has its own median. For the lower-half data set, the median is the average of its sixth item, 24, and its seventh item, 25. This value, 24.5, is called the **lower quartile** for the original data set. For the upper-half data set, the sixth and seventh items are both 32, so this is the median for this part of the data. This value is called the **upper quartile** for the original data set.

The lower quartile, the median, and the upper quartile are often represented by the symbols Q1, Q2, and Q3, because they represent the values *one-quarter* of the way, *two-quarters* of the way, and *three-quarters* of the way from the bottom to the top of the list of data items.

These three values separate the data into four sections, as shown here.

22, 23, 23, 23, 24, 24   25, 25, 25, 26, 27, 27

28   29, 30, 30, 31, 31, 32   32, 32, 32, 33, 35, 39

continued ▶

Because the full data set has an odd number of items, the middle item, 28, is the median and is shown separately, not in any of the four sections. The lower quartile, 24.5, is simply the number halfway between 24 and 25 and is not in the list. The upper half of the data set, from 29 through 39, happens to have values of 32 as its sixth and seventh items, so the upper quartile is 32.

The Box-and-Whiskers Plot

One standard way to represent quartiles and the four sections of a data set graphically is a **box-and-whiskers plot** (or **box plot,** for short). A box plot shows the data set in relation to a number line.

The box part of the diagram goes from Q1 to Q3 and roughly represents the middle 50% of the data. The numbers in this group range from 25 through 32. The vertical line through the box is at Q2.

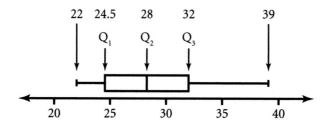

The lines extending from the box—called the *whiskers*—go from the box out to the lowest and highest values. The left whisker represents the lowest 25 percent of the data, which covers data items from 22 through 24. The right whisker represents the highest 25 percent of the data and includes data items from 32 through 39.

The box plot makes it easy to see that the highest item in the data set is farther from the median than is the lowest item.

Data Pictures

Choose one or more data sets, either from your work in this unit or from some other source, and represent the data using each of these three methods.

- A frequency bar graph
- A stem-and-leaf plot
- A box-and-whiskers plot

Then compare the results.

- How are the representations similar?
- How are the representations different?
- What advantages does each method have?

Sample frequency bar graph

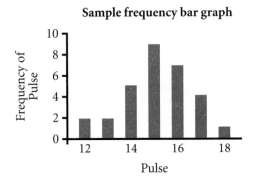

Sample stem-and-leaf plot

Stem	Leaf
2	8
3	3, 5
4	1, 2, 7
5	1, 3, 3, 4, 6, 9

Sample box-and-whisker plot

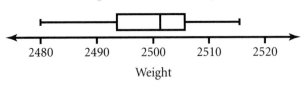

More Knights Switching

If you liked *POW 9: The Big Knight Switch,* here is a similar problem that is a bit more challenging.

This time there are three white and three black knights instead of two and two. They are sitting on a board with four rows and three columns.

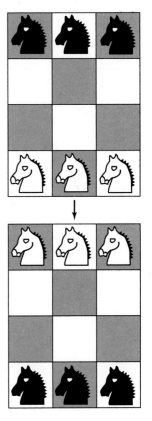

The knights start in the positions shown in the top diagram. The goal is for the three black knights to change places with the three white knights, so that they end up as shown in the bottom diagram.

The knights are allowed to make only the moves allowed in chess.

1. Can the knights switch places?

2. If so, what is the least number of moves it will take them to switch? How do you know your answer is the least?

3. If it is not possible, explain why you are sure it is not.

A Knight Goes Traveling

If you like playing around with knight moves, here's one more problem for you.

In *POW 9: The Big Knight Switch,* you worked with four knights on a 3-by-3 chessboard. In *More Knights Switching,* you worked with six knights on a 4-by-3 chessboard.

Now look at a single knight. Imagine placing that knight in one corner of a square chessboard of some unknown size, as shown.

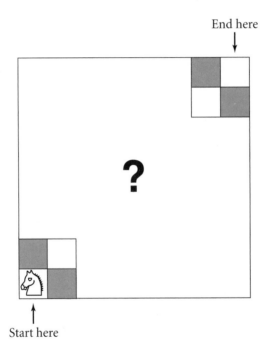

End here

?

Start here

How many moves will it take for the knight to get to the diagonally opposite corner?

The answer depends on the size of the chessboard. It may even be impossible for some cases.

Begin with a 2-by-2 board, then look at a 3-by-3 board, and gradually work your way up. Keep track of the fewest number of moves in each case. Look for patterns.

Data for Kai and Mai

In *The Best Spread,* you were asked to make up two data sets, set X and set Y, that together fit these two conditions.

- Set X was to have a larger standard deviation than set Y.
- Set Y was to have a larger range than set X.

In other words, based on standard deviation, set X was to be more spread out, but based on Tai's method, set Y was to be more spread out.

In this activity, your task is to find two other pairs of data sets. In each case, deciding which set is more spread out will depend on the method used for measuring data spread.

1. Make up a pair of data sets, R and S, so that both of these statements are true.
   - Based on standard deviation, set R is more spread out than set S.
   - Based on Kai's method, set S is more spread out than set R.

     Reminder: Kai measures the spread of a set of data by finding the distance from each number to the mean and then adding those distances.

2. Make up another pair of data sets, U and V, so that both of these statements are true.
   - Based on standard deviation, set U is more spread out than set V.
   - Based on Mai's method, set V is more spread out than set U.

     Reminder: Mai measures the spread of a set of data by ignoring the highest and lowest data items, and finding the largest distance from any remaining data item to the mean.

Making Better Friends

In *Making Friends with Standard Deviation,* you explored what happened to the mean and the standard deviation of a set of data when you added the same number to each member of the set or multiplied each member of the set by the same number.

In this activity, your task is to explore some other questions about mean and standard deviation. Here are some ideas to start you off.

1. How can you add new data items to a set so that you don't change the mean or the standard deviation?

2. How can you add new data items to a set so that you can keep the mean the same but make the standard deviation as large as you like?

Don't restrict yourself to these two questions. Make up some questions of your own and explore them. Report on everything you have discovered.

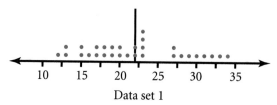

Data set 1

Mean is 22.

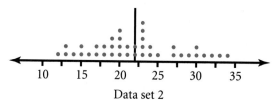

Data set 2

Mean is 22.

Mean Standard Dice

You know that, in the long run, a normal die will come up 1 one-sixth of the time, 2 one-sixth of the time, and so on. Thus, for example, with 36 rolls, you should expect about six 1s, six 2s, and so on.

1. Find the mean and standard deviation for the data represented by this ideal result for 36 rolls (six 1s, six 2s, and so on). Explain the computations you used.

2. Suppose the die was rolled 360 times, with the ideal result of sixty 1s, sixty 2s, and so on.

   a. Write down a guess about what the mean and the standard deviation would be for this set of data. Explain your guess.

   b. Actually find the mean and the standard deviation for this situation. Explain your computations.

   c. Are the actual mean and the actual standard deviation the same as in Question 1? Explain why you think they came out the same or different.

3. Suppose you used a pair of fair dice, rolling them together and each time finding the sum of the numbers on the two dice.

   a. What would be the ideal result for these sums if the dice were rolled 36 times? That is, how many 12s would you get? How many 11s? Continue to find the number for each sum.

   b. Write down a guess about what the mean and the standard deviation would be for this ideal result of 36 sums for a pair of dice. Explain your guess.

   c. Find the actual mean for this ideal result. How does it compare with your mean in Question 1?

   d. Find the actual standard deviation for this ideal result. How does it compare with your standard deviation in Question 1?

More About Soft Drinks, Calculators, and Tests

1. The soft drink company from *Can Your Calculator Pass This Soft Drink Test?* has a new shape for its soda bottles. Here are the key facts.

   - The mean amount of soda put in each bottle is 1000 mL.

   - The standard deviation for the amount in each bottle is 5 mL.

   These new bottles won't spill unless the filling machine puts more than 1010 mL into a bottle. They will still appear to be less than full if they contain less than 995 mL.

   What percentage of bottles should be removed from the assembly line, either because they have a risk of spill or because they appear not to be full?

2. The calculator manufacturer from that same assignment has improved the durability of her product. The time until repair now averages 1154 days, with a standard deviation of only 36 days.

   Suppose the manufacturer replaces any calculator needing repair in fewer than 3 years. Estimate the percentage of the calculators she would need to replace.

3. Big State University has decided that a good score on a college entrance examination should not, by itself, be a guarantee of admission. However, they do want a score of at least 400. The test still has a mean score of 490 and a standard deviation of 120.

   Estimate the percentage of students who will be disqualified because of low test scores.

Are You Ambidextrous?

In this activity, you will compare the reflexes of your two hands.

Have a partner hold a ruler vertically between your thumb and forefinger, so that the lower end of the ruler is level with your fingers. Spread your thumb and forefinger as wide apart as possible. Your partner should hold the ruler from its upper end.

Your partner will say "Drop" at the moment he or she drops the ruler. As the ruler falls past your fingers, you try to pinch it as quickly as you can.

1. Do the experiment with one of your hands 20 times, each time recording the place where you grab the ruler.

2. Find the mean and the standard deviation of your data.

3. Draw a normal curve, with a horizontal scale, that has the same mean and standard deviation as your data. Show where the standard deviation marks are located.

4. Now do the ruler experiment, this time with your other hand.

5. How many standard deviations from the mean was the experiment result from your other hand? What percentage of the time do you think you would get a result that far from the mean using your first hand? Explain.

6. Do you think you are ambidextrous? Why or why not? If you don't know what the word *ambidextrous* means, ask someone or look it up in a dictionary.

Family of Curves

You've seen that functions based on similar algebraic expressions often have similar graphs. For example, certain equations have graphs that are lines.

In this activity, your goal is to explore families of curves.

Start with a simple function, such as $y = x^2$. Carefully examine how its graph changes as you make various changes in the equation.

Make a poster showing your results.

More Height and Weight

In the supplemental activity *Height and Weight,* you did a preliminary investigation of the relationship between the height and the weight of objects.

In this activity, you will follow up with a more detailed investigation that looks specifically at the relationship between height and weight for people.

1. Find out the heights and weights for various people. Try to use individuals with a wide range of different heights, from small children to adults.

2. Make an In-Out table of your data, using height as the *In* and weight as the *Out*. Then make a graph of your data.

3. Try to find a formula that comes close to fitting your data.

4. Use what you've learned to estimate how much a 10-foot-tall person might weigh. Explain your reasoning.

Out of Action

The general manager of the Slamajamas basketball team has a difficult decision to make. A key player, Marcus Dunkalot, suffered a sprained knee on March 12, 5 weeks before the playoffs, and was put on the disabled list. The playoffs begin on April 18.

It is now April 7, and the general manager has just received the physical therapist's report, shown here. Based on this report, the manager needs to decide immediately whether to keep Marcus on the disabled list. If he does, it will be for the remainder of the regular season, so Marcus will be disqualified for the playoffs.

PROFESSIONAL PHYSICAL THERAPY

Patient's name: Marcus Dunkalot

Sex: Male   Height: 6'8"

Age: 24   Weight: 225 lb

Diagnosis: sprained knee

Prescribed treatment: strengthen and stretch

3/12	Mr. Dunkalot was administered a Cybex strength test upon arrival. Quadriceps of the injured leg measured 55 foot-pounds in extension. Normal measurement for a player to return to play without reinjury is 250 foot-pounds.
3/17	Daily regimen is contributing to patient's progress. Cybex test measures 85 foot-pounds.
3/24	Some swelling earlier in the week. General reports of less pain. Cybex test measures 135 foot-pounds.
3/28	Less swelling. Range of motion has shown marked increase. Cybex test measures 175 foot-pounds.
4/2	Leg strength continues to improve, but more slowly, as expected. Cybex test measures 195 foot-pounds.
4/7	Progress continues. Cybex test measures 210 foot-pounds.

---

continued ▶

Here are the advantages and disadvantages of each choice.

- If he takes Marcus off the disabled list now, he can hope that Marcus will be well in time for the playoffs. But if Marcus is not ready in time, then the Slamajamas will have one less player available for the rest of the season, including the playoffs.

- If he keeps Marcus on the disabled list, he can sign another player (of lesser ability) to take his place. But then he gives up all hope of having Marcus for the playoffs.

1. Graph Marcus Dunkalot's progress.

2. Imagine that you are the general manager. The team owners want a complete report on why and how you made your decision. What will you decide? Write the report.

Gettin' on Down to One

This problem is about a certain pair of rules for generating a sequence of numbers.

You can start with any positive whole number (1, 2, 3, 4, and so on) as the first term. After that, you find each term by applying one of two rules to the current term. The decision about which rule to apply depends on whether the current term is an odd number or an even number.

- If the current term is an odd number, the rule is to multiply the current term by 3 and then add 1 to get the next term in the sequence.

- If the current term is an even number, the rule is to divide the current term by 2 to get the next term in the sequence.

For example, suppose the starting number is 7. Because this is an odd number, you use the first rule: Multiply by 3 and then add 1. That gives 22, so the second term in the sequence is 22. Because 22 is even, you now use the second rule: Divide by 2. That gives 11, so the third term in the sequence is 11.

Following the correct rule each time, starting with 7, you generate the sequence 7, 22, 11, 34, 17, 52, 26, 13, 40, 20, 10, 5, 16, 8, 4, 2, 1, 4, 2, 1, 4, 2, 1, 4, 2, 1, ...

continued ▸

Notice that any time this procedure gets to 1, the sequence will then go 4, 2, 1 again, over and over. Therefore, you should consider the sequence to be finished if it reaches 1.

When you start with 7, it takes 16 steps to reach 1. The starting number itself is not counted as a step.

1. Use the pair of rules to generate and record sequences for each of these starting numbers.

    a. 6

    b. 9

    c. 21

    d. 33

2. In Questions 1a through 1d, you should have found that each sequence eventually reached 1. For each case, find out how many steps it took to reach 1. The only starting number that gets to 1 in only one step is 2, and the only starting number that gets to 1 in only two steps is 4.

3. Which starting numbers will get down to 1 after only three steps? Four steps? Five steps? Explore.

4. Describe a way to find starting numbers that will produce very long sequences, such as 100 steps.

5. What other observations can you make about how this procedure works?

Programming Down to One

In *Gettin' on Down to One,* you examined a process for generating sequences of numbers. Here's a summary of that process.

You begin with a starting number. You find each number after that by applying one of two rules to the current number. The decision about which rule to apply depends on whether the current number is odd or even.

- If the current number is odd, the rule is to multiply the current number by 3 and then add 1 to get the next number.
- If the current number is even, the rule is to divide the current number by 2 to get the next number.

Your task in this activity is to write a program that will generate the sequence for you.

There are several types of programs that you could write. Here are three possibilities.

Option 1

The program asks the user for a number and then tells the user what the next number is.

For example, the screen for such a program might look like the first display.

The user of the program enters the number 13. Once the program responds, the user enters 40. To generate the whole sequence, the user needs to enter each term and have the calculator generate the next term.

> Give me a number
> ? 13
> The next number
> in the sequence is 40

Option 2

The program asks the user for a number and then generates a sequence of terms based on the two rules. For example, the screen for such a program might look like this display.

> Give me a number
> ? 13
> The sequence goes
> 13, 40, 20, 10, 5,
> 16, 8, 4, 2, 1

continued ▶

Everything except the number 13 following the question mark is done by the program.

Option 3

The program asks the user for a number and then tells the user how many steps it takes to get to 1, without actually showing the terms. For example, the screen for such a program might look like this display. Again, everything except the number 13 following the question mark is done by the program.

```
Give me a number
? 13
The sequence gets
to 1 in 9 steps
```

Whatever options you work on, your program will have to determine which of the two rules to use, depending on whether a given number is odd or even.

You may want to consult a manual to find out how to display words on the screen, how to put the decision about which rule to use into your program, and so on.

Shadows

Similar Triangles and Proportional Reasoning

Shadows—Similar Triangles and Proportional Reasoning

What Is a Shadow?

This unit asks, "How long is a shadow?" But before you can answer this question, you need to ask some others: "What is a shadow?" "Where do shadows come from?" and "Are there different kinds of shadows?"

You will also need to do some experimenting. You will identify the variables that affect the length of a shadow in a certain situation. You'll then design and carry out some experiments to relate those variables to the shadow's length.

In-Out tables will help you keep track of data from shadow experiments. They will also be useful as you explore cutting a pie in the first Problem of the Week.

The Pit and the Pendulum and *Shadows* units each start with identifying variables and discovering their relationships to each other. But each has its own set of mathematical ideas for you to learn, and the two units are really quite different.

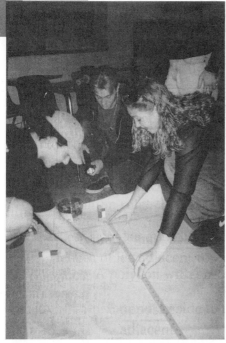

Cody Boling, Ethan Fitzhenry, and Nikki Robinson measure the length of the shadow in the "Shadow Data Gathering" experiment.

How Long Is a Shadow?

As you walk around, I stay with you.

I'm sometimes ahead of you and sometimes behind you.

If you were smaller, I would be too.

If you come toward me, I often get bigger (especially at night).

If you shine a light on me, I disappear.

What am I?

You probably guessed the answer to the simple riddle. But there are some riddles about shadows that are not so clear. The big riddle for this unit is: How long is a shadow?

It's easy to answer by saying, "It depends." But what does it depend on? Here are three shadow situations for you to think about.

- You are walking on the sidewalk at night. A streetlamp is behind you. You look down at your shadow on the sidewalk. As you get farther away from the streetlamp, what happens to your shadow's length?

- You put your hand in front of a projector lamp to make a "rabbit" shadow. As you move your hand farther away from the lamp, how does the shadow on the screen change?

- On a sunny afternoon, you jog outside around a track. You look down and see your shadow on the track. As you run, how does your shadow change?

Each situation has different variables that affect the size and location of the shadow.

What Is a Shadow?

As a first step toward answering the unit question, discuss these questions.

- What exactly is a shadow?

- What are some different situations in which shadows are formed?

- Why might some shadows change size as you move while others do not?

continued ▶

Defining Variables

We'll begin by focusing on "lamp shadows." These shadows are formed on the ground by a lamp or flashlight located above and behind the object.

If you walk near a lighted streetlamp, your body casts a shadow on the ground.

The length of your shadow depends on several other lengths, distances, or angles that might change. The lengths, distances, and angles that can change are the variables for the situation.

Make a written list of the variables for the lamp-shadow situation. Be specific in describing your variables. For example, if you think changes in a certain length or distance might be important, state exactly what that distance measures—from where to where.

Preparing for *Experimenting with Shadows*

As a group, choose one of the variables that can be measured that you think affects the shadow's length. Write a careful plan for a set of experiments concerning this variable. The purpose of the experiments is to explore whether your variable really affects the length of a shadow and, if so, how changing the value of the variable affects shadow length.

continued ▶

Here are some ideas to keep in mind as your group puts together the plan.

- Change only the variable that you are investigating, keeping everything else fixed. Your plan should state which variables you are keeping fixed as well as the one you are changing.

- Think about the materials you'll use in your experiment. For example, what will be your light source? What kind of object will cast the shadow? Include that information in your plan.

- Think about how you will change the value of the variable you are studying. You should try at least five different numerical values for this variable. For each value, you should measure the length of the shadow.

Experimenting with Shadows

In *How Long Is a Shadow?* your group planned an experiment to investigate how a particular variable might affect the length of a shadow. Your assignment is to carry out this experiment.

1. As closely as you can, carry out your plan. Record your results. You may need to change the plan once you actually try to carry it out.

2. Describe the results of your experiment, giving all the numerical data you collected. Include a labeled picture or diagram of your experiment. Note the specific values you used for the variables that you kept fixed.

3. Write about what you learned. What effect did the variable have on the length of shadows? State your conclusions clearly. Use the data you collected as evidence for your conclusions.

Attach a copy of the plan set up by your group.

The Shadow Model

In Question 2 of *Experimenting with Shadows,* you made a picture or diagram of your experiment. Your next task is to refine this diagram to create a **mathematical model** for the shadows problem—that is, create a mathematical representation that contains the key features of the experimental situation.

Your experiments will serve as a guide to understanding the model. The correctness of the model depends on whether it is confirmed by experimental evidence. Using the model will allow you to apply some powerful geometric principles. Ultimately, you will use the model to create a solution to the general shadow problem.

1. Based on class discussions so far, list the key variables that determine the length of a lamp shadow.

2. Create and label a new diagram that shows these key variables.

3. State the assumptions that you are making in using this diagram to represent the shadow problem.

Cutting the Pie

You can sometimes organize information from experiments into In-Out tables to help you understand what's happening in the experiment.

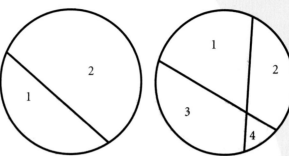

The diagrams show the results of a pie-cutting experiment. In the first picture, one cut across the pie has created two pieces. In the next, two cuts have created a total of four pieces.

Notice that the cuts do not necessarily go through the center of the pie, but they do have to be straight and go all the way across the pie. In addition, the pieces do not have to be the same size or shape.

The two diagrams show different possible results from making three cuts in the pie.

In the first case, the three cuts produced six pieces. In the second case, the three cuts produced seven pieces. It's also possible to produce five or even only four pieces from three cuts.

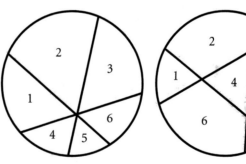
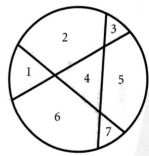

You should be able to convince yourself that seven is the largest number of pieces that can be created by three cuts across the pie. The purpose of the pie-cutting experiment is to answer this question.

What is the largest number of pieces that can be produced from a given number of cuts?

continued ▶

The information from the diagrams has been organized into the In-Out table.

Number of cuts	Maximum number of pieces
1	2
2	4
3	7
4	?
5	?

Your task is to extend and analyze this table.

1. Find the largest number of pieces you can get from four cuts and from five cuts. Put those numbers into the table.

2. a. Try to find a pattern describing what is happening in the table.

   b. Use your pattern to find the largest possible number of pieces from ten cuts.

   c. Try to explain *why* this pattern is occurring.

3. Try to find a rule for the In-Out table. If you used a variable for the input, how could you write the output as a formula in terms of that variable?

○ *Write-up*

1. *Problem Statement*

2. *Process:* Include important diagrams you used while working on this problem.

3. *Solution:* Include these things.
   - Your In-Out table as far as you took it
   - Any patterns you found in the table, expressed in words, in symbols, or both
   - Your answer for the largest possible number of pieces after ten cuts
   - Explanations for your answers

4. *Extensions*

5. *Self-assessment*

Poetical Science

Augusta Ada Byron (1815–1852) was the daughter of the famous English poet George Gordon Byron (Lord Byron) and a mathematician, Anne Isabella Milbanke Byron. Augusta Ada Byron is known today as Ada Lovelace, having married Lord Lovelace.

At the time she lived, people thought there was a sharp separation between the skills needed for the sciences and the abilities involved in creative arts like poetry. This conflict between science and poetry goes back at least to the Greek philosopher Plato. He was suspicious of poetry and felt that "it gives no truth of its own, stirs up the emotions, and thereby blinds mankind to the real truth."

Ada Byron Lovelace. Illustration courtesy of Stock Montage, Chicago, Illinois.

Ada Lovelace was unusual for her time, perhaps because of her parents' combination of interests. Her formal education was traditional in its emphasis on facts, but her father's imaginative influence kept appearing. Her teacher thought that studying mathematics would cure Ada of being too imaginative, but Ada responded that she had to use her imagination in order to understand mathematics.

Lovelace was herself a mathematics teacher and encouraged her students to use their imaginations. She emphasized metaphors and visual images. Her students used colored pens, rulers, and compasses (which were then considered "vulgar instruments") to make drawings that would help their understanding.

continued ▶

She wrote:

> Imagination is the *Discovering* Faculty preeminently. It is that which penetrates into the unseen worlds around us, the worlds of Science. It is that which feels and discovers what *is*, the REAL which we see not, which *exists* not for our *senses*.

At the age of 18, Ada met Charles Babbage, who was then in the process of inventing what is now considered the world's first computer. Working with him was the realization of a dream, because it allowed her to combine her imagination with her analytical skills. She was able to work on mathematics as a joy, not as the duty it had been when her mother made her study it.

Lovelace played an important role in the development of computer science by explaining Babbage's inventions to a larger audience. In her written *Notes*, she created a unified vision of their usefulness, combining metaphors about the creative power of the machines with detailed technical descriptions of their operation.

In honor of these contributions, a computer language was named after her. The language Ada is one of the most sophisticated tools for studying artificial intelligence and is used by the military.

Your Assignment

1. The article talks about the use of imagination in understanding mathematics. Think about two situations this year when your imagination helped you in this class. Write about those experiences.

2. Do you agree or disagree with what Plato said about poetry? Write what you think about poetry and truth.

This article is adapted from the preface "Poetical Science" in the book *Ada, The Enchantress of Numbers*, edited by Betty Alexandra Toole (Mill Valley, CA: Strawberry Press, 1992). The book contains Ada's letters and her description of the first computer.

Shadow Data Gathering

In this activity and in *Working with Shadow Data,* you will gather data and then look for a function that fits the data.

You have seen that the length of a shadow seems to depend on three variables.

- The height of the light source (L)
- The distance from the object to the light source (D)
- The height of the object (H)

Using S for the length of the shadow, the first central problem of the unit, finding the length of a lamp shadow, might be stated in this way.

> *What formula can be used to express S as a function of L, D, and H?*
> *That is, how can you get a formula for a function f so that S = f (L, D, H)?*

Your Task

In *Experimenting with Shadows,* you described the general effects of these three variables on a lamp shadow. In this activity, you will describe these effects more precisely, in quantitative terms. Your task is to focus on just one of these three variables, keeping the other two fixed. If your chosen variable is H, for example, you will pick specific values for L and D and not change them. You will then gather data about how S changes as you change H.

Organize your data into an In-Out table in which the input is the value of your chosen variable and the output is the length of the shadow.

Your group will prepare a report on its work. Include these things.

- Your In-Out table, in a form appropriate for display
- Identification of your chosen variable
- A diagram of your experiment, including the fixed values used in your experiments for the other two variables

An *N*-by-*N* Window

The unit's central problem is to find a formula relating shadow length to certain other variables. This activity involves finding a formula, but the setting—window frames—is completely different.

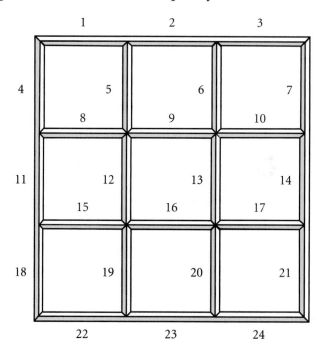

The window frame is made of wood strips that separate the glass panes. Each glass pane is a square that is 1 foot wide and 1 foot tall. As the numbering shows, it would take 24 feet of wood strips to build a frame for a window 3 feet by 3 feet.

Your task is to develop a formula for the total length of wood strip needed to build square windows of different sizes.

You may want to do this by gathering data and making an In-Out table from different examples. Or you may prefer to study the window, looking for insights that lead directly to a formula. Or you may do a combination of both. In either case, generalize the problem to a window *N* feet by *N* feet. As usual, include all your drawings, In-Out tables, and graphs.

Working with Shadow Data

You now have an In-Out table that shows data relating one of the shadow variables, *L*, *D*, or *H*, to the shadow length, *S*.

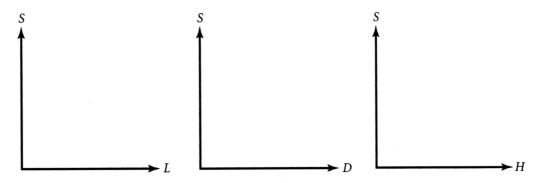

1. Your next task is to use your In-Out table to examine the relationship between the shadow length and your chosen shadow variable.

   • Graph your data using suitable scales.

   • Examine the pairs of numbers in your table and look for patterns.

   Develop one or more general statements about the relationship between the shadow length and your variable. These statements need not involve equations or rules.

2. When you finish working with your group's data, study the data gathered by another group.

More About Windows

In *An N-by-N Window,* you investigated the length of wood strip needed for square window frames. You found a general formula for the amount of wood strip needed for a square window frame of any size.

Generalize to a rectangular window frame. Find a formula in terms of *M* and *N* for the amount of wood strip needed for the frame of an *M*-by-*N* window.

You may want to gather data about a variety of examples, put the data in an In-Out table, and look for a pattern. Then look for an algebraic rule that describes your table.

Or you may prefer a more analytic approach. You might reason through why such a window frame should use a particular amount of wood strip. If you find a rule this way, verify it using some examples.

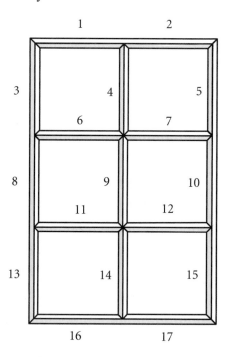

The Shape of It

You've done some measurements and experiments involving shadows. Now it's time to try a different approach.

Beginning with the first activities, you're going to leave the world of shadows and enter the more abstract realm of geometry. Angles, polygons, lengths of sides—these will be the focus of your attention for much of the unit. Eventually, these ideas will fit together to answer the shadowy questions that are lurking in the background.

For now, your focus will be on issues that arise when you represent a real-world situation with a mathematical diagram. How can you be sure that your diagram on paper is true to the full-size reality? What does it mean to have a diagram that is the same shape as the real thing?

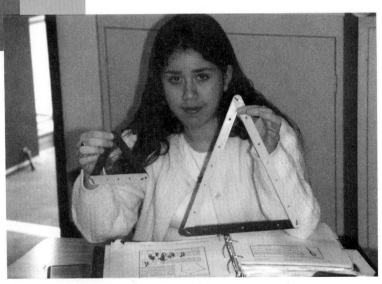

Jackie Hernandez shows the class possible counterexamples.

Draw the Same Shape

1. Draw a small, simple picture on a sheet of grid paper. Then draw another version of your picture that is exactly the same shape as the first one. Your second drawing should be larger than the original. You have to decide what "exactly the same shape" means to you.

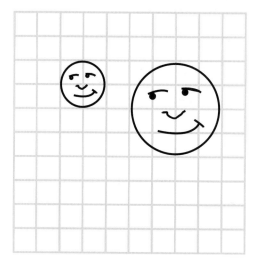

2. Renata was making illustrations for a book on basic drawing techniques. She came up with this figure as the first step for drawing a house.

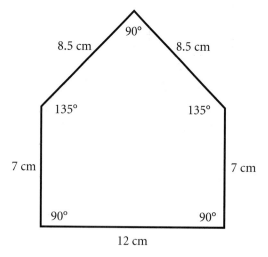

continued ▶

Kim decided he liked this shape, but he needed a bigger version of the house for his work.

a. Give a set of length and angle measurements that you think Kim might be able to use in place of those shown in the diagram. State clearly which measurements change and which, if any, do not.

b. Carefully draw a diagram that has your suggested measurements and compare it with Renata's. Does it have the same shape?

3. Consider the following pairs of figures. In each case, state whether you consider them to be the same shape or not. Explain why.

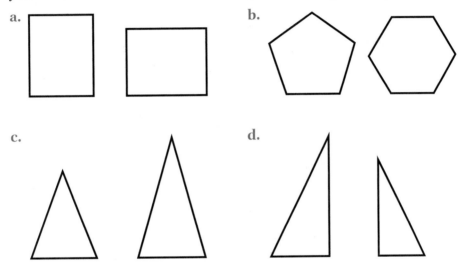

a.

b.

c.

d.

4. Based on your experience with Questions 1 through 3, write your ideas in response to this question.

How can you create a diagram that has exactly the same shape as a given one?

How to Shrink It?

Lola, Lily, and Lulu love Renata's house, but they find it a little too large for their liking.

They want to shrink the house to a smaller size while keeping it exactly the same shape.

After a long discussion, each came up with a strategy for drawing the smaller house.

Lola's way: Keep all the angles as they are, and subtract 5 centimeters from the length of each side.

Lily's way: Keep all the lengths as they are, and divide all the angles by 2.

Lulu's way: Keep all the angles as they are, and divide the lengths of all the sides by 2.

Shrink the house by using each method. Show what result each method produces. Explain why each method does or does not work.

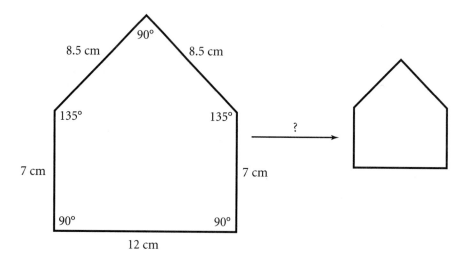

The Statue of Liberty's Nose

The Statue of Liberty in New York City has a nose that is 4 feet 6 inches long.* What is the approximate length of one of her arms?

1. Solve the problem. You might want to think about your own nose and arms.

2. Pick two other body measurements and find the approximate length that these measurements should be on the Statue of Liberty.

3. Examine what you did with the three examples from Questions 1 and 2. How was your work the same in the three cases? How did it change from case to case?

4. State how this problem is similar to the problem of drawing a house that has the same shape as another house.

5. What connection do you see between this problem and the shadow problem?

*Measurement taken from *How They Built the Statue of Liberty* by Mary Shapiro (Random House, 1985).

Make It Similar

You have seen that two polygons are called **similar** if their corresponding angles are equal and their corresponding sides are proportional.

The phrase "corresponding sides are **proportional**" means that if you compare each side of the first polygon with the corresponding side in the second polygon, the **ratios** of those corresponding lengths are all equal.

Here's a problem where you need to think carefully about which sides are corresponding.

First, suppose you have one triangle with sides that have lengths of 2 inches, 3 inches, and 4 inches.

Next, suppose there is a second triangle that is known to be similar to the first one.

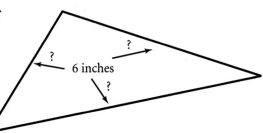

Suppose you know that one of the sides of this larger triangle is 6 inches long.

Unfortunately, you don't know which side of the large triangle has this length.

What can you say about the lengths of the other sides of the larger triangle? Try to find all the possibilities. There's more than one answer. Don't assume that the triangles shown are drawn to scale.

Pool Pockets

Imagine a modified pool table in which the only pockets are those in the four corners. The diagram shows such a table as viewed from above.

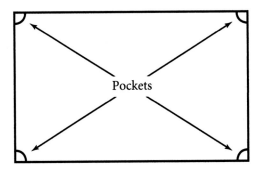

Pockets

This POW will use the view from above all the time, with different parts of the table labeled as in the next diagram.

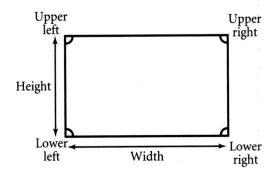

Upper left

Upper right

Height

Lower left

Width

Lower right

Imagine that a ball is hit from the lower-left corner in a diagonal direction that forms a 45° angle with the sides.

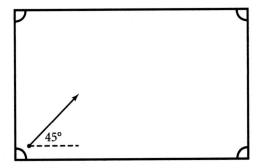

45°

continued ◗

Imagine that every time the ball hits an edge of the table, it bounces off, again at a 45° angle, and that it continues this way until it hits one of the corner pockets perfectly. For example, using the previous diagram, the first few bounces of the ball would look like this.

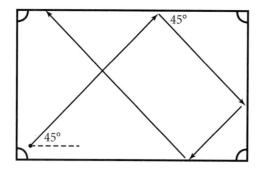

Of course, the path of the ball will depend on the table's shape. For example, with a square table, the ball would go directly into the opposite corner without bouncing at all.

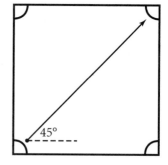

Your task in this POW is to investigate what happens to the ball and how this depends on the dimensions of the table.

You should make these assumptions.

• Both the height and the width of the table are whole-number distances.

• The ball is always shot at an angle of 45°.

• The ball is always shot from the lower-left corner.

You might consider these questions.

• Does the ball always hit a pocket eventually?

• If so, which pocket does it hit?

• If it does hit a pocket, how many times does it bounce before it hits the pocket?

You may have some other questions of your own.

continued ▶

You will probably find it very useful to have grid paper for your investigation. Use the width of the squares on the grid paper as your unit of length. In that way, you'll be able to track more easily the ball's exact path.

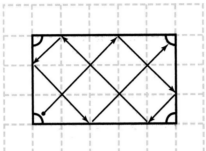

For example, the diagram illustrates how you might use grid paper to show the path of the ball for a 3-by-5 table. The dotted lines represent the lines of the grid paper, and the solid rectangle is the outline of the table. The diagonal lines with arrows show the path of the ball. At the end of the path, the ball is about to go into the pocket in the upper-right corner of the table.

Happy bouncing!

○ *Write-up*

1. *Subject of Exploration:* Describe the subject that you are investigating. What questions do you want to explore?

2. *Information Gathering:* Basing your comments on your notes (which should be included with your write-up), state what happened in the specific cases you examined.

3. *Conclusions, Explanations, and Conjectures:* Describe any general observations you made or conclusions you reached. Wherever possible, explain why the particular conclusions are true—that is, try to prove your general statements. But also include conjectures, or statements that you only think are true.

4. *Extensions:* What questions do you have that you were not able to answer? What other investigations would you do if you had more time?

5. *Self-assessment*

A Few Special Bounces

POW 14: Pool Pockets asked you to consider all possible tables with whole-number dimensions. That's a lot of cases! This activity will get you started with a simpler version.

Specifically, you will consider all tables that have a height of exactly 2 units. Remember that the width is supposed to be a whole number.

The first three such tables are shown. The path of the ball has been drawn for the first of these tables.

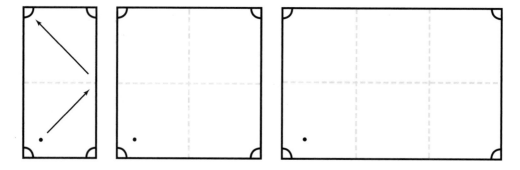

Be sure to consider not just these three, but all tables with a height of 2 units.

What happens to the ball on these tables? Consider the specific questions posed in *POW 14: Pool Pockets*, and ask some others of your own.

Ins and Outs of Proportion

The two triangles are similar because the ratios between a side of Triangle 1 and the corresponding side of Triangle 2 are equal for all three sides. In other words, the three ratios—$\frac{12}{9}$, $\frac{20}{15}$, and $\frac{24}{18}$—are equal. For triangles, having corresponding sides **proportional** guarantees similarity.

What about ratios of sides within a triangle? For example, the ratio of the longest to the shortest side in Triangle 1 is $\frac{24}{12}$. The corresponding ratio in Triangle 2 is $\frac{18}{9}$. These two ratios are also equal, because both fractions are equal to 2.

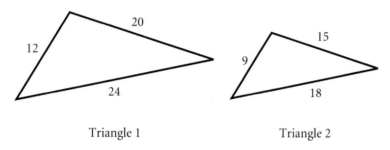

Triangle 1   Triangle 2

1. Compare the other ratios within each triangle. Find a different ratio of two sides of Triangle 1, and then compare that with the corresponding ratio in Triangle 2. Do this for all possible pairs of sides. What do you conclude?

2. Form another pair of similar triangles. To get corresponding sides in proportion, you can use any convenient set of lengths for the sides of the first triangle. Then multiply those numbers by a fixed value to get the lengths of the sides of the second triangle.

   Repeat Question 1 for your new pair of triangles.

3. Form a third pair of triangles, but this time make them not similar. Examine the ratios of sides within one triangle compared with the ratios of sides within the other. What do you conclude?

continued ▶

4. Now consider these two triangles. Assume that they are similar. The ratios $\frac{r}{x}$, $\frac{s}{y}$, and $\frac{t}{z}$ are all equal.

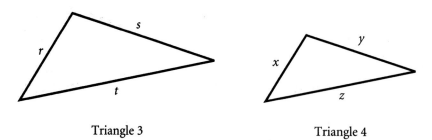

Triangle 3   Triangle 4

a. Based on your results in Questions 1 through 3, identify a pair of ratios—one using sides of Triangle 3 and one using sides of Triangle 4—that you think are equal.

b. Find as many pairs of equal ratios as you can.

Similar Problems

In each of the four pairs of polygons, assume that the second polygon is similar to the first. Also assume for each pair that the two similar figures have the same orientation—that is, corresponding sides are facing the same way.

For each pair of figures, perform these steps.

- Set up equations to find the lengths of any sides labeled by variables.
- Find the length that solves each equation.

Note: Measuring the diagrams will probably not give correct answers because the diagrams may not be drawn exactly to scale.

1.

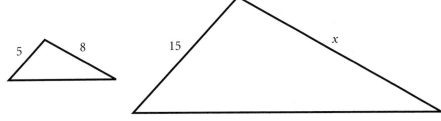

2.

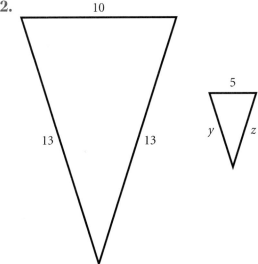

continued ▶

3.

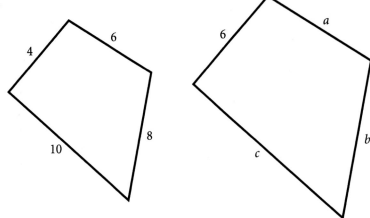

4.

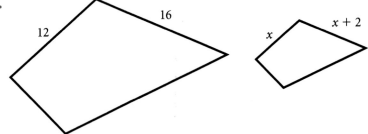

Inventing Rules

In working with similar triangles, you often have to solve equations involving proportions.

Suppose one triangle has sides of lengths 6, 9, and 14. Suppose there is a similar triangle with shortest side of length 15. To find the longest side of the second triangle, represent it with x and find the value of x that satisfies this equation.

$$\frac{6}{15} = \frac{14}{x}$$

This is one of several possible equations for x.

Some equations of this kind are easier to solve than others. Sometimes the particular numbers involved suggest shortcuts that make them easy to solve.

In each equation, the letter x stands for an unknown number. Use any method you like to find what the number x stands for. Write down exactly how you do it.

Be sure to check your answers.

1. $\frac{x}{5} = 7$ 2. $\frac{x}{6} = \frac{72}{24}$ 3. $\frac{x}{8} = \frac{11}{4}$

4. $\frac{x}{7} = \frac{5}{3}$ 5. $\frac{x + 1}{3} = \frac{4}{6}$ 6. $\frac{5}{13} = \frac{19}{x}$

7. $\frac{2}{x} = 6$ 8. $\frac{9}{x} = \frac{x}{16}$

9. For each ratio in Question 4 and 8, draw a pair of similar triangles with side lengths that would create the ratios in the proportion.

10. Use any method you wish to solve for the unknown number in this proportion. Draw a pair of similar triangles that would reflect this proportion.

$$\frac{x + 3}{7} = \frac{x}{3}$$

Polygon Equations

As in *Similar Problems,* assume that in each pair of figures, the second polygon is similar to the first, with the same orientation. Perform these steps.

- Set up equations to find the lengths of the sides labeled by variables.
- Solve each equation and find the labeled lengths.
- Find any remaining lengths of either figure.

Note: These diagrams are not necessarily drawn to scale.

1.

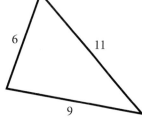

2.

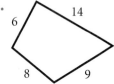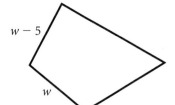

3.

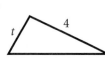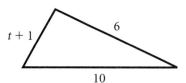

Triangles Galore

Triangles and other polygons play a vital role in geometry. What statements about polygons are true in every case? Are some true only about triangles? Are some types of triangles more special than others?

As you explore properties and relationships, counterexamples and generalizations will be important tools.

You've seen that angles are an important component of what makes polygons similar. How much do you really need to know about the angles of a polygon to be sure that the polygons have the same shape? Can those angles be anything you like?

In this part of the unit, you will focus your efforts on developing principles that will allow you to prove that two shapes are similar.

Nader Al-Ammari investigates properties of triangles.

Triangles Versus Other Polygons

Triangles are special polygons. Some of them have special names. A triangle with at least two sides of equal length is called **isosceles.** If all three sides have the same length, the triangle is called **equilateral.** An equilateral triangle is a special kind of isosceles triangle.

Triangles have some characteristics that not all polygons share.

Each pair of statements includes one version about triangles and another version about polygons in general.

For each pair, first decide whether the statement is true for triangles. Then try to find a counterexample that shows the statement is not true for polygons in general.

Statements 1

If two angles in one triangle equal two angles in another triangle, then their third angles must be equal.

If two angles in one polygon equal two angles in another polygon, then their other angles must be equal.

Statements 2

If two triangles have their corresponding angles equal, then the triangles are similar.

If two polygons have their corresponding angles equal, then the polygons are similar.

Statements 3

If two triangles have their corresponding sides proportional, then the triangles are similar.

If two polygons have their corresponding sides proportional, then the polygons are similar.

Statements 4

Every triangle with two equal sides also has two equal angles.

Every polygon with two equal sides also has two equal angles.

Angles and Counterexamples

Equations with Angles

Based on the angle-sum formulas you have developed, write and solve
equations to find the angles of each polygon. *Note:* The figures may not
be drawn to scale.

1.

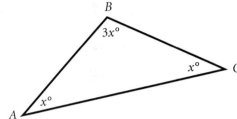

2.

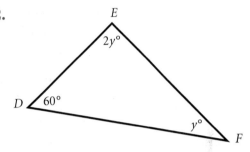

3.

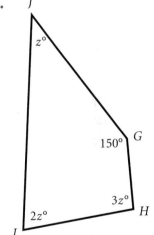

4.

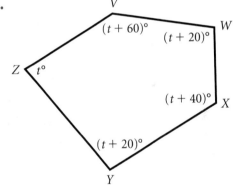

continued ▶

More Counterexamples

Attempt to draw counterexamples for each statement. Be sure to show all your work, including appropriate diagrams.

If you don't think a counterexample exists, then explain why you think there isn't one.

5. If two triangles are both isosceles, then the triangles are similar.

6. If two triangles have two pairs of corresponding sides proportional, then the triangles are similar.

7. Every triangle with two equal angles also has two equal sides.

Why Are Triangles Special?

You have seen that triangles seem to be different from other polygons with regard to similarity.

In this activity, you will investigate why triangles are special.

1. Pick four lengths and form a quadrilateral using those lengths for the sides.

   Then try to use the same four lengths to form a quadrilateral that is not similar to the first. Is this possible?

2. Repeat Question 1, starting with more than four lengths. That is, pick some lengths and form a polygon using those lengths. Using the same lengths, try to form a polygon that is not similar to the first. Is this possible?

3. Start with three lengths and use them to form a triangle. As in Questions 1 and 2, try to use the same lengths to form a triangle that is not similar to the first. Is this possible?

More Similar Triangles

Your work with triangles will often involve a situation in which the similar triangles are overlapping, or inside one another. Here is an example of overlapping similar triangles. Do you see the two similar triangles?

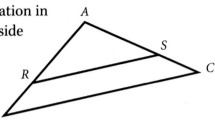

If you were to separate them and redraw them, the two triangles would look like this.

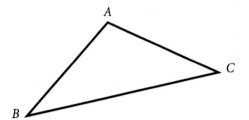

Find the unknown lengths in these pairs of overlapping similar triangles.

1.

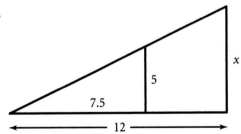

2.

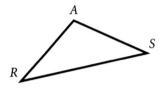

Use your calculator to solve these proportions for the unknown side length.

3. $\dfrac{4.6}{27.6} = \dfrac{2.5}{x}$

4. $\dfrac{301}{d} = \dfrac{426}{5964}$

Are Angles Enough?

You've seen that the lengths of the sides of a triangle determine the triangle. Here is another way to say this.

> If the lengths of the three sides of one triangle are the same as the lengths of the three sides of a second triangle, then the two triangles are **congruent.**

What about angles? Do the angles of a triangle determine the triangle? What happens if two triangles have the same three angles?

1. Start with angles of 40°, 60°, and 80°. Each group member should draw a triangle using these three angles.

   Did the triangles all come out congruent? Were they all similar? Why or why not?

2. Do the same thing as in Question 1, but start with a different set of angles. You might draw an arbitrary triangle and then have each group member use the same angles as that triangle.

   Did the triangles all come out congruent? Were they all similar? Why or why not?

3. Go back to the angles of 40°, 60°, and 80° from Question 1. This time, decide as a group on the length each group member will use for the side connecting the angles of 40° and 60°. As before, have each group member draw a triangle using angles 40°, 60°, and 80° but also using the given length in the given position.

   Did the triangles all come out congruent? Were they all similar? Why or why not?

In Proportion

If two polygons are similar, then their sides must be proportional. That is, the ratio of the lengths of corresponding sides must be constant. For instance, if one polygon is similar to another, the sides of one might all be twice as long as the sides of the other, or all three times as long, and so on.

The corresponding angles of the two similar figures are the same, because angles do not increase proportionately the way lengths do.

Proportionality occurs often in real life. It's important to know what should change and what should not. Consider each situation.

1. A recipe for baking 30 brownies includes these details.
   - 4 ounces of chocolate
   - A 350° oven
   - 2 cups of sugar
   - 1 cup of flour
   - A 9-inch-by-13-inch-by-2-inch pan
   - 25 minutes baking time

continued ▶

Suppose you want to use a similar recipe to make 60 brownies—twice as many. Which numbers listed in this recipe would you double? Which would you change in some other way? Which numbers would you keep as they are?

2. A 200-mile car trip might involve driving for 4 hours at an average speed of 50 miles per hour.

   If you needed to go 600 miles, under similar driving conditions, which numbers, if any, would change proportionately?

3. The seniors at Central High School are planning a dance. They are expecting 100 students to attend. Here are some numbers they have developed in connection with their initial plans.

   • The planning committee will have 5 members.

   • Tickets will be $12 each.

   • The dance will last 3 hours.

   • They will have $600 worth of refreshments.

   • They will hire two different bands.

   Before the committee can get started, the dance suddenly becomes "the event of the season." Now 250 students are expected. Which numbers from the plan should change proportionately? How might other numbers change? Explain your thinking.

What's Possible?

You have found that the three angles of a triangle are subject to a very strict condition—their sum must be exactly 180°. This activity poses a related question about the lengths of the sides of a triangle.

Can any three numbers be the lengths of the sides of a triangle?

You may want to work on this question using physical materials, such as straws, paper strips, or pieces of spaghetti.

Experiment to find the answer to the question. You might start with the numbers 2, 3, and 4. Can you draw a triangle that has sides with these lengths? Be sure to choose a unit of length.

Try three other values, such as 3, 6, and 11. Keep making up numbers and testing them. Keep track of which sets of lengths are possible and which are not. What conclusions can you reach about the three sides of a triangle?

What About Other Polygons?

What principles can you formulate about the lengths of the sides of a quadrilateral? Play around with examples, using physical materials if you choose.

Continue with polygons that have more sides. Can you state a general principle for whether a given set of numbers can be used as sides of a polygon?

Very Special Triangles

You saw in *Why Are Triangles Special?* that triangles are a special category of polygons. Some triangles are even more special than others.

You're familiar with the isosceles triangle, in which at least two sides have the same length, and the equilateral triangle, in which all three sides are equal in length.

Another special category is the **right triangle.** A right triangle has one angle that measures 90°.

1. Why must the other two angles of a right triangle be **acute**—that is, less than 90°?

Triangle *ABC* is a right triangle with a right angle at vertex *C*. The small square inside that vertex is a standard symbol that indicates a right angle.

The two sides of the triangle that form the right angle, $\overline{AC}$ and $\overline{BC}$, are called the **legs** of the triangle. The third side (opposite the right angle), $\overline{AB}$, is called the **hypotenuse.** These two terms apply only to right triangles, not to triangles in general.

Each acute angle of a right triangle is formed by the hypotenuse and one of the legs. For example, $\angle A$ is formed by the hypotenuse, $\overline{AB}$, and by the leg $\overline{AC}$. The leg that helps form an acute angle is said to be **adjacent to** that angle.

For example, $\overline{AC}$ is the leg (or side) adjacent to $\angle A$. That same leg is said to be **opposite** the other acute angle. For example, $\overline{AC}$ is the leg opposite $\angle B$.

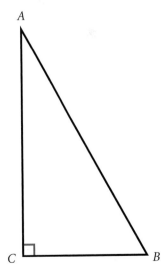

continued ▸

2. What statements can you make about how the lengths of the sides of a right triangle compare to each other? Explain your reasoning.

3. Draw a right triangle, measure the legs, and then draw a right triangle with legs twice as long as those of your first triangle.

   a. How does the hypotenuse of the new triangle compare to the hypotenuse of the original?

   b. How do the acute angles of the new triangle compare to the acute angles of the original?

   c. What do your answers to parts a and b tell you about the two triangles?

4. Draw a right triangle in which the acute angles are different sizes. Is the longer leg opposite or adjacent to the larger of the acute angles? Do you think this is true for all right triangles?

5. Is it possible for a right triangle to be isosceles? Equilateral? Explain your answers.

Angle Observations

You've seen that angles play a very important role in similarity. It's useful to know when two angles are sure to be equal.

The diagram is made up of three line segments: $\overline{AG}$, $\overline{AH}$, and $\overline{BE}$.

Segment BE intersects $\overline{AG}$ at point C and intersects $\overline{AH}$ at point D.

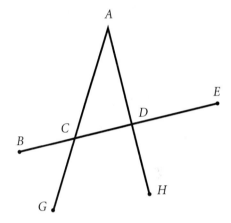

There are many angles in the diagram. Your task is to explore the relationships among those angles and then test whether your observations apply to similar diagrams.

Whatever method you use to measure the angles, look for explanations for why certain angle relationships hold. That will allow you to generalize your results.

1. Create a diagram that is roughly like the one above. Label your diagram to match the diagram here.

2. Measure each angle in your diagram and record your results. Then answer these questions.

   • Which angles are equal to which other angles?

   • What angle-sum relationships can you find?

3. Create other diagrams like this one, or compare your results with those of other group members. Test whether the relationships you found in Question 2 were examples of a general principle.

4. Generalize from your work in Questions 2 and 3. Justify your general conclusions.

More About Angles

A line that intersects two or more other lines is called a **transversal** for those other lines. For example, in this diagram, the line labeled ℓ is a transversal for the pair of lines m and n.

The case where lines m and n are **parallel** is especially important and is the subject of this activity.

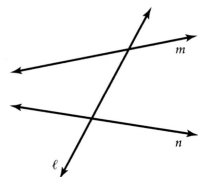

Note: The mathematical symbol for parallel is $\|$. We read "$m \| n$" as "m is parallel to n."

In the diagram, $\overleftrightarrow{BD}$ and $\overleftrightarrow{EG}$ are parallel lines. In symbols, $\overleftrightarrow{BD} \| \overleftrightarrow{EG}$.

$\overleftrightarrow{AH}$ is a transversal that intersects $\overleftrightarrow{BD}$ at C and intersects $\overleftrightarrow{EG}$ at F.

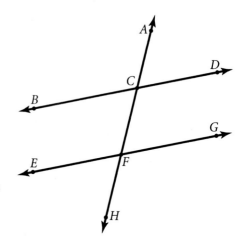

1. Create a diagram like the one here and label your diagram to match. Be sure that $\overleftrightarrow{BD}$ and $\overleftrightarrow{EG}$ are parallel.

2. Measure each angle in your diagram and record your results. Then answer these questions.

   • Which angles have the same measure as which other angles?

   • What angle-sum relationships can you find?

   Pay special attention to the angle relationships that occur because $\overleftrightarrow{BD}$ and $\overleftrightarrow{EG}$ are parallel.

3. Create other diagrams like this one. Test whether the relationships you found in Question 2 were examples of general principles.

4. Generalize from your work in Questions 2 and 3, and justify your general conclusions.

Trying Triangles

Suppose you have a straight pipe cleaner of a given length. For convenience, label the two ends of the pipe cleaner *A* and *C*.

Suppose that the pipe cleaner is bent into two portions, with one portion twice as long as the other. Label the place where the bend is made B, as shown, so that the segment from *B* to *C* is twice as long as the segment from *A* to *B*.

Choose a fourth point, *X*, at random somewhere on the longer section. The pipe cleaner is bent at that place as well. Now the pipe cleaner might look like this.

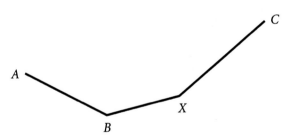

Can the three segments of the pipe cleaner be made into a triangle by changing the angles at the two bends? As you might suppose, the answer depends on the location of point *X*.

So here's the question.

> *If point X is chosen at random along the section from B to C, what is the probability that the three segments of the pipe cleaner can be made into a triangle?*

"Chosen at random" means that all positions between *B* and *C* are equally likely to be selected as the location for point *X*.

continued ▶

As part of your investigation, you will probably first need to come up with the answer to this question.

Suppose three lengths, a, b, and c, are given. What condition must these lengths satisfy for it to be possible to make a triangle with sides of these three lengths?

Note: You may find it helpful to assume that the original pipe cleaner has a specific length.

○ *Write-up*

1. *Problem Statement*

2. *Process*

3. *Solution:* Your solution should include answers to both questions posed.

4. *Extensions*

5. *Self-assessment*

Adapted from *Mathematics Teacher* (November 1989), National Council of Teachers of Mathematics, volume 82, number 6.

Inside Similarity

How do you make small triangles inside larger ones so that the small ones are similar to the large ones?

The diagram shows three congruent triangles. In the bottom two triangles, a dotted line segment has been drawn that connects two sides of the triangle and cuts off a smaller (shaded) triangle.

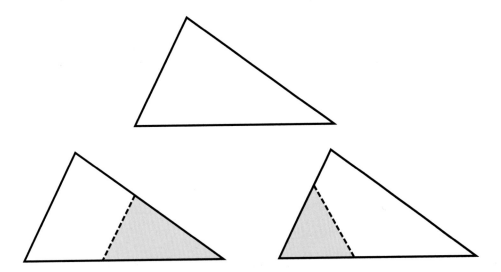

In the first case, the smaller triangle appears to be similar to the larger one. In the second case, the smaller triangle seems not to be similar to the larger one.

Your task is to investigate and report on the difference between these two cases. If you connect points on two sides of a triangle, when does the smaller triangle created in this way come out similar to the original?

You may want to begin your investigation by tracing the original triangle. Then experiment by drawing some segments on your tracing. Find as many ways as you can to draw lines that cut off small triangles similar to the original triangle.

Describe in words those segments that can be used to cut off a small triangle that is similar to the larger one, and explain your answer.

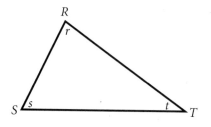

A Parallel Proof

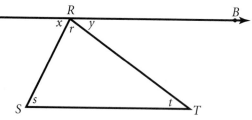

You've seen, based on measurement estimates, that the sum of the angles of any triangle seems to be 180°. Now it's time for you to prove that this must be true.

In this activity, you will be given a diagram involving an arbitrary triangle. Your task is to show how to use this diagram to prove the angle sum property for triangles.

To get started, suppose *RST* is any triangle.

For convenience, the angles of this triangle are labeled *r*, *s*, and *t*.

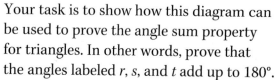

Draw a line through vertex *R* that is parallel to the opposite side. This is the line *AB* shown in the diagram, in which $\overleftrightarrow{AB} \parallel \overleftrightarrow{ST}$. The angles *ARS* and *BRT* have been labeled *x* and *y*.

Your task is to show how this diagram can be used to prove the angle sum property for triangles. In other words, prove that the angles labeled *r*, *s*, and *t* add up to 180°.

You may find it helpful to extend the sides of the original triangle.

In this diagram, both $\overleftrightarrow{EF}$ and $\overleftrightarrow{GH}$ are transversals for the parallel lines *AB* and *CD*. Think about how to use the transversals to get information about angles.

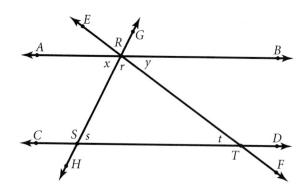

Angles, Angles, Angles

Find the designated angles in each diagram. Lines marked with arrows are parallel.

1.

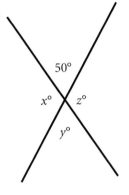

2.

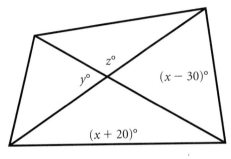

3.

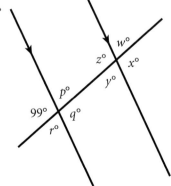

4.

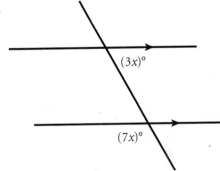

5.

6.

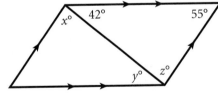

The Lamp Shadow

You're about to solve the lamp shadow problem. As a result, you will get a rather nice formula expressing the length of the shadow in terms of other variables.

As you prepare to complete this challenge, this part of the unit returns to applications of triangles and similarity. These applications will lead you to develop more equations, and you'll expand the range of equations that you know how to solve.

You'll also have a chance to explore spiralaterals in the final POW of the unit.

Gabriela Asuncion, Daisy Corona, Brenda Jimenez, and Geno Bernal revisit the Lamp Shadow experiment.

Bouncing Light

If you take a flashlight and shine it into a mirror, the light will "bounce off" the mirror. In this activity, you will look at the angles involved in such bouncing light.

Set up your flashlight and mirror in a way similar to that shown in the sketch. One person can hold the mirror, another can shine the flashlight at the mirror, and a third can mark the path along which the reflection of the light leaves the mirror.

The angle between the mirror and the incoming ray of light from the flashlight is called the *angle of approach*. The angle between the mirror and the ray bouncing off the mirror is called the *angle of departure*.

1. Use a protractor to measure the angle of approach and the angle of departure. You may want to trace the path of the flashlight's beam to and from the mirror onto chart paper.

   What do you notice about the relationship between the two angles?

2. Repeat this experiment, but change the angle at which the beam goes into the mirror. Do this two more times, each time with different angles. Does the relationship that you observed between the angles always seem to hold true? Write down what you've noticed about the relationship.

3. Hold the mirror in different positions to look at other things in the classroom. What do you notice about the position of the mirror? Write about the relationship between your observations in Question 2 and how you have to hold the mirror. Try to explain what looking at something in a mirror has to do with bouncing light.

Now You See It, Now You Don't

Use the principle of light reflection and your protractor to investigate the following problems. You will need to trace each diagram in this assignment.

1. A person is standing at point *A* and is looking toward the mirror. Which letters of the alphabet can this person see?

A
•   *B C D E F G H I J K L M*

━━━━
Mirror

2. Two spiders are on opposite walls. A large mirror is placed on the floor.

Copy the diagram, and show exactly where on the floor the spiders should look to see each other. Can you find two triangles that are similar in your drawing? Why must they be similar?

Mirror Magic

You can actually use a mirror and the principle of light reflection to measure the heights of objects. All that is needed is a flat surface on which to place your mirror.

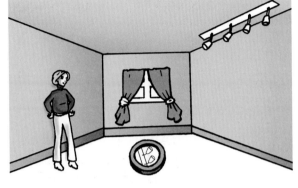

The method uses similar triangles, and it begins like this.

Place the mirror on the ground some distance from the object. Move slowly backward, away from the object, while looking down at the mirror. At some point, you should be able to see the object in your mirror.

Measure some things that are easier to measure than the height of the given object, and then apply geometry.

1. Use this method to find the height of an object in the classroom.

2. Make a diagram that shows your method. Label the distances that you measured. Show how you used these measurements to find the height of the object.

3. Assign variables to the distances you measured. Set up an In-Out table in which these variables are labels for the input columns. Assign a variable for the height of your object, and use that as the label for the output column. Enter your actual measurements as the first row of inputs in your table. Put the height you got as the first output.

4. Make up some numbers for your input variables. Based on your made-up numbers, find the new height of the object. In other words, find some more rows for your table, calculating the output (the height) in terms of the numbers you made up. The geometric situation in your diagram should serve as the basis for your calculation.

5. Give a description in words or write an equation that shows the relationship between the height of the object and your input variables.

Mirror Madness

A family of spiders has found a bunch of mirrors on the ground. The spiders have been positioning themselves to see one another in the mirrors.

They are dangling in the order shown, although their distances from one another, their heights off the ground, and the positions of the mirrors are not necessarily drawn to scale.

Sister spider, who is 48 inches off the ground, can see Momma spider in a mirror that is on the ground between them. This mirror is 20 inches from the point directly below Sister spider and 30 inches from the point directly below Momma.

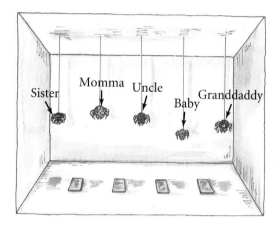

Momma spider can see Uncle spider in a mirror that is 10 inches from the point below Momma and 5 inches from the point below Uncle.

Uncle spider can see Baby spider in a mirror that is 8 inches from the point below Uncle and 6 inches from the point below Baby.

Finally, Baby spider can see Granddaddy spider in a mirror that is 12 inches from the point below Baby and 16 inches from the point below Granddaddy.

Your Task

Find the height of each spider. (You already know the height of Sister spider.) Show your equations and how you solved them.

A Shadow of a Doubt

By now, you have probably developed a diagram similar to this one to represent the lamp shadow problem.

In terms of this diagram, the length of a shadow (S) depends on three things: the height of the light source (L), the height of the object casting the shadow (H), and the distance along the ground from that object to the light source (D).

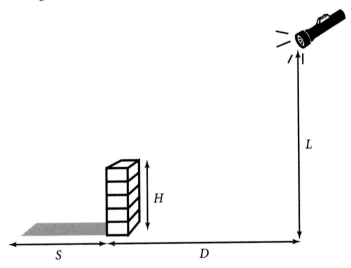

1. What triangles do you see in the diagram? Which of them are similar? Why must these triangles be similar?

2. Use your knowledge of similar triangles to write an equation that expresses a relationship among the four variables.

3. Here is some information about a specific situation involving Shoshana, a lamppost, and her shadow. Shoshana is standing 9 feet from the lamppost. The light at the top of the lamppost is 12 feet high, and Shoshana is 5 feet tall. Use your equation from Question 2 to find the length of Shoshana's shadow.

4. Find the length of the shadow if $L = 11$, $H = 6$, and $D = 13$.

5. Write a description in words of how to find the length of a shadow when L, H, and D are given.

To Measure a Tree

To find the height of a tree, you might climb to the top and drop a long tape measure to the ground while still holding the start of the tape. You'd need a friend on the ground to read off the tree's measurement.

Although straightforward, this method has many difficulties and dangers. Fortunately, there are less hazardous methods, including some that use similar triangles. Your task is to use your knowledge of similar triangles to invent methods for measuring a tree. Use the illustration for ideas.

Create at least three methods, relying on different pairs of similar triangles, to find the height of the tree. Clearly identify the similar triangles in each method. Explain how these triangles are used in your method. You may have to draw extra lines in your diagram to illustrate the situation.

Write down what you'd have to know in each situation and how you'd use that information to figure out the height of the tree.

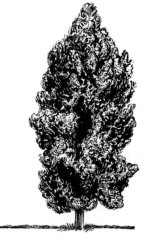

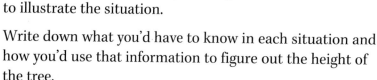

Spiralaterals

A *spiralateral* is a sequence of line segments that forms a spiral-like shape. It is easiest to draw spiralaterals on grid paper.

Each spiralateral is based on a sequence of numbers. To give an example, let's suppose your sequence is 3, 2, 4. To draw the spiralateral, you need to choose a starting point. The starting direction is always toward the top of the paper.

The first number is 3, so the spiralateral begins with a segment that goes up 3 spaces, as in diagram A.

Before each new number, the spiralateral turns clockwise 90° (that is, a quarter turn to the right). So the second segment of the spiralateral will go to the right. The second number is 2, so the spiralateral moves 2 spaces in that direction. This gives diagram B.

The spiralateral again turns clockwise 90°, so it's now facing "down." This time it goes 4 spaces, because the third number in the sequence is 4, giving diagram C.

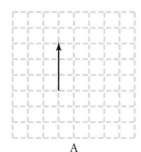
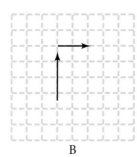
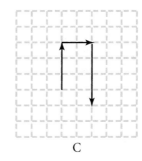

A	B	C

You've now reached the end of the original three-number sequence. But this is where it gets interesting. Instead of stopping, you go back to the beginning of the sequence and continue where you left off.

Make another 90° clockwise turn (so you're now going to the left), and move 3 spaces, because the first number of the sequence is 3. Then make another 90° turn and go up 2 spaces, another 90° turn and 4 spaces to the right, another 90° turn and 3 spaces down, and so on. Continue until you get back to the place where you started (if you ever do!). It's as if your sequence, instead of just being 3, 2, 4, is 3, 2, 4, 3, 2, 4, 3, 2, 4, 3, 2, 4, ...

continued ▶

In this example, the diagram will look like this after ten steps.

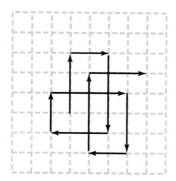

As you can see, in two more steps, you'll be back to the start. The complete spiralateral for this sequence looks like this.

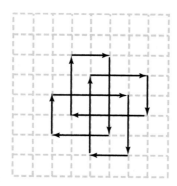

For this POW, you will explore the idea of spiralaterals. Here are some things you can do as part of your exploration.

- Make some spiralaterals, using your own sequences of numbers.
- Look for patterns.
- Make up questions about spiralaterals.
- Try sequences of different lengths.
- Make up new rules.

Whenever you come to a conclusion about spiralaterals, try to explain why your conclusion is true.

Save your notes as you work on the problem. You will need them to do your write-up.

continued ▶

○ *Write-up*

1. *Problem Statement:* Explain what a spiralateral is and how it is formed.

2. *Process*

3. *Results and Conclusions*

    a. Show the results of some of the specific examples you investigated. Include diagrams as appropriate.

    b. What patterns did you notice among your results in part a? Summarize what you concluded from your examples, stating your conclusions clearly.

    c. Justify your conclusions as fully as you can. In other words, for any patterns that you found in your examples, explain why you believe that those patterns hold true in general.

    d. What questions that you have not discussed occurred to you as you worked on this problem?

4. *Extensions*

5. *Self-assessment*

More Triangles for Shadows

In *A Shadow of a Doubt*, you used a diagram like this to write an equation relating the four variables *S*, *H*, *L*, and *D*.

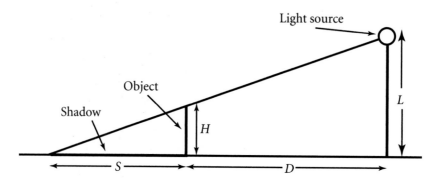

Your goal now is to develop an algebraic expression for S in terms of *L*, *H*, and *D*. In other words, you need to find a function *f* so that $S = f(L, D, H)$.

You may have used similar triangles to develop an equation involving these four variables. For example, if you compare the small triangle in the diagram to the large triangle, you might get this equation.

$$\frac{S}{S+D} = \frac{H}{L}$$

The fact that S appears in two places in this equation complicates the process of using it to get a general expression for S in terms of the other variables.

Your task is to find another way to use similar triangles so that using proportions gives you a simpler initial equation. See if you can find a pair of similar triangles in the diagram and write an equation for proportionality in which the variable S appears only once. Once you get that simpler equation, use it to find the expression for S in terms of *L*, *H*, and *D*.

The Sun Shadow

Now that you've created a function to determine shadow length for the lamp shadow situation, here comes another problem—the sun shadow. This new situation will lead you into the world of trigonometry.

As you solve the sun shadow problem, you will be finishing the unit—and Year 1 as well. Your portfolio for this unit asks you to look back over the entire year.

Keahara Monroe prepares to study sun shadow problems.

The Sun Shadow Problem

At night or if you are indoors, shadows are likely to be caused by lamps or streetlights. But during the day, the shadows you see outdoors usually come from blocking off the light from the sun.

For such a "sun shadow," it doesn't make much sense to talk about the height of the light source off the ground or the distance along the ground from the light source to the object casting the shadow. The variables L and D are rather meaningless for sun shadows.

So what does the length of a sun shadow depend on? What are the relevant variables? At what time of day would a shadow be the longest, and when would it be the shortest? Why?

Speculate on these questions and any others of your own. Think about how you might go about expressing the length of a shadow in terms of the appropriate variables. Draw and label a picture or diagram that illustrates your thinking.

Write down any conjectures or ideas you have.

Right Triangle Ratios

Many of the problems you've worked on, including the basic diagram for the shadow problems, have involved right angles.

You've also seen that ideas of similarity involve ratios of sides of triangles.

So it's natural to think about ratios of sides within right triangles.

1. Carefully draw a right triangle *ABC* with a right angle at *C* and with a 55° angle at *A*. Make your triangle larger than the one shown here so any measurement error will be less significant.

    Record the lengths of the three sides of your triangle.

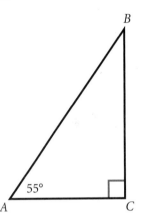

In *Very Special Triangles,* you were introduced to this terminology.

- Side $\overline{AB}$ is called *the hypotenuse* of this right triangle.
- Side $\overline{BC}$ is called *the leg opposite angle A.*
- Side $\overline{AC}$ is called *the leg adjacent to angle A.*

2. Find each of the ratios. If your measurements in Question 1 involved fractions of a unit, convert your fractions to decimals and use those decimal values to compute the ratios.

    a. $\dfrac{\text{length of leg opposite angle } A}{\text{length of leg adjacent to angle } A}$

    b. $\dfrac{\text{length of leg opposite angle } A}{\text{length of hypotenuse}}$

    c. $\dfrac{\text{length of leg adjacent to angle } A}{\text{length of hypotenuse}}$

3. Do you think your classmates will get the same results for Questions 1 and 2 that you got? Explain in detail why or why not.

Sin, Cos, and Tan Revealed

Did you ever wonder what those keys on your calculator that say "sin," "cos," and "tan" are all about? Well, here's where you find out.

You've seen that whenever two right triangles have another angle in common, the triangles must be similar, and so the corresponding ratios of lengths of sides in those triangles are equal.

These ratios depend only on that common acute angle. Each ratio of lengths in the right triangle has a name. The study and use of these ratios is part of a branch of mathematics called **trigonometry.**

Suppose you are given an acute angle—that is, an angle between 0 and 90 degrees. You can create a right triangle in which one of the acute angles is equal to that given angle. Suppose you label that triangle as shown in the diagram, so that $\angle A$ is equal to the acute angle you started with.

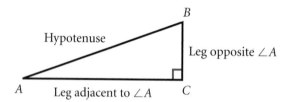

The trigonometric ratios are then defined as explained on the following pages. The principles of similarity guarantee that these ratios will be the same for every right triangle that has an acute angle the same size as $\angle A$.

Sine of an Angle

The **sine** of $\angle A$ is the ratio of the length of the leg opposite $\angle A$ to the length of the hypotenuse. The sine of $\angle A$ is abbreviated as **sin A.** For example, in $\triangle RST$, the leg opposite $\angle R$ has length 4, and the hypotenuse has length 7, so $\sin R = \frac{4}{7}$.

continued ▶

In summary

$$\sin A = \frac{\text{length of leg opposite } \angle A}{\text{length of hypotenuse}}$$

Or simply,

$$\sin A = \frac{\text{opposite}}{\text{hypotenuse}}$$

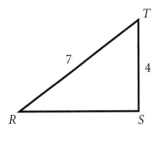

Cosine of an Angle

The **cosine** of $\angle A$ is the ratio of the length of the leg adjacent to $\angle A$ to the length of the hypotenuse. The cosine of $\angle A$ is abbreviated as cos A. For example, in $\triangle UVW$, the leg adjacent to $\angle U$ has length 3, and the hypotenuse has length 5, so $\cos U = \frac{3}{5}$.

In summary

$$\cos A = \frac{\text{length of leg adjacent to } \angle A}{\text{length of hypotenuse}}$$

Or simply,

$$\cos A = \frac{\text{adjacent}}{\text{hypotenuse}}$$

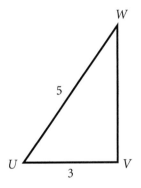

Tangent of an Angle

The **tangent** of $\angle A$ is the ratio of the length of the leg opposite $\angle A$ to the length of the leg adjacent to $\angle A$. The tangent of $\angle A$ is abbreviated as tan A. For example, in $\triangle HKL$, the leg opposite $\angle H$ has length 2, and the leg adjacent to $\angle H$ has length 6, so $\tan H = \frac{2}{6}$.

In summary

$$\tan A = \frac{\text{length of leg opposite } \angle A}{\text{length of leg adjacent to } \angle A}$$

Or simply,

$$\tan A = \frac{\text{opposite}}{\text{adjacent}}$$

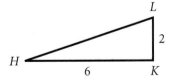

continued ▶

Trigonometric Functions on a Calculator

Any scientific calculator or graphing calculator has keys that will give you the values of these functions for any angle.

In some calculators, you enter the size of the angle and then push the appropriate trigonometric ratio key. For other calculators, you do the opposite.

You have been measuring angles using degrees as the unit of measurement, but there are other units for measuring angles. Most calculators that work with trigonometric functions have a mode key that you can set to "deg."

Homemade Trig Tables

For any acute angle, all right triangles containing that angle will have equal trigonometric ratios for that angle, because the triangles will all be similar. It will be useful to compile a table of the ratios for common angles. You can use it in future work when you are given only one angle of a right triangle with missing information. One way to create this information and explore this property of trigonometry is to draw some triangles, measure the sides, and find the ratios.

To get results that are reasonably accurate, you should make all the sides of your triangles at least 5 centimeters long and measure each length to the nearest millimeter.

Your teacher will assign you one or more angles to investigate. For each angle you are assigned, do these three things.

• Draw a right triangle using that angle as one of the acute angles.

• Measure the length of all three sides.

• Compute the appropriate ratios to find the sine, cosine, and tangent of the assigned angle.

Compute your ratios to the nearest hundredth. You will combine your results with those of your classmates to create a table of trigonometric values.

Your Opposite Is My Adjacent

Every right triangle has two acute angles. You can learn some interesting facts about trigonometry by using both of them.

Use the labeling in $\triangle ABC$ to answer the questions.

1. What relationship must exist between angles A and B?

2. Express the ratio $\frac{BC}{AB}$ in two ways.

   a. As the tangent, sine, or cosine of $\angle A$

   b. As the tangent, sine, or cosine of $\angle B$

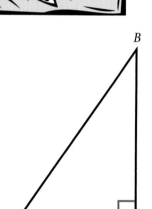

3. Use your results from Questions 1 and 2.

   a. Write a general formula for the sine of an angle as the cosine of a related angle.

   b. Write a similar general formula for the cosine of an angle as the sine of a related angle.

The Tree and the Pendulum

1. Now that you have been introduced to trigonometry, it's time to look again at how to measure the height of a tree.

    Here are the key facts.

    • Woody is 12 feet from the tree.

    • Woody's line of sight to the top of the tree is at an angle of 70° up from horizontal.

    • Woody's eyes are 5 feet off the ground.

    Describe how Woody could find the height of the tree using trigonometry and these measurements.

2. You can apply trigonometry to the situation from *The Pit and the Pendulum*.

    Suppose a 30-foot pendulum has an initial amplitude of 30°.

    How far is the bob from the center line when the pendulum starts? In other words, what is the distance labeled d?

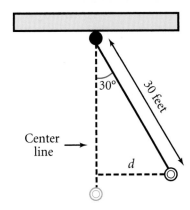

Sparky and the Dude

1. Sparky the Bear

Sparky the Bear is atop a 100-foot tower. He is looking out over a fairly level area for careless people who might start fires. Suddenly, he sees a fire starting. He marks down the direction of the fire. But he also needs to know how far away from the tower the fire is.

To figure out this distance, Sparky grabs his handy protractor. Because he is high up on top of the tower, he has to look slightly downward toward the fire. He finds that his line of sight to the fire is at an angle of 6° below horizontal. *Note:* This diagram is not to scale.

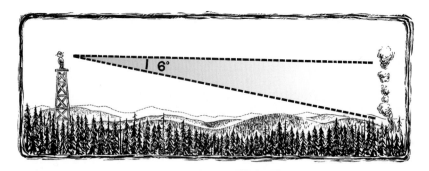

How far is the base of Sparky's tower from the fire?

2. Dude on a Cliff

Shredding Charlene is out surfing. She catches the eye of her friend Dave the Dude, who is standing at the top of a cliff. The angle formed by Charlene's line of sight and the horizontal measures 28°.

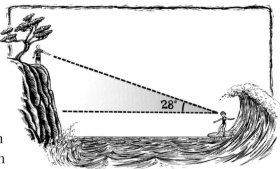

Charlene is 50 meters from the bottom of the cliff. Charlene and Dave are both 1.7 meters tall. They are both 16 years old. The surfboard is level with the base of the cliff.

How high is the cliff?

A Bright, Sunny Day

Suppose you are standing outdoors on a bright, sunny day.

You look up toward the sun and estimate the sun's **angle of elevation.** In other words, you measure the angle shown as θ, the Greek letter *theta.*

How can you find the length of your shadow (S) using this angle and your own height (H)?

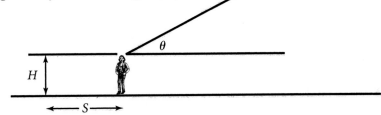

Beginning Portfolio Selection

As the first step in assembling your portfolio for *Shadows,* look back over the main unit problem.

One of the key ideas in this unit is the concept of similarity.

1. Explain what similarity means. Give an intuitive description and then use the formal definition.

2. Choose two or three activities in the unit that helped you understand and use this concept. Explain why you chose those activities and how they contributed to your ideas about similarity.

Christina Pickett prepares work for her portfolio.

Shadows Portfolio

Now that *Shadows* is completed, it is time to put together your portfolio for the unit. In addition to writing a cover letter and selecting papers, your portfolio work for *Shadows* includes looking back over the entire first year of your IMP experience.

Cover Letter

Look back over *Shadows* and describe the central problem of the unit and the main mathematical ideas. Your description should give an overview of how the key ideas were developed and how they were used to solve the central problem.

As part of your portfolio, you will be selecting some activities that you think were important in developing the key ideas of this unit. Your cover letter should include an explanation of why you select the particular items you do.

Selecting Papers

Your portfolio for *Shadows* should contain these items.

• *Beginning Portfolio Selection*

Include the activities from the unit that you selected in *Beginning Portfolio Selection,* along with your written work about the concept of similarity and about the activities.

• Activities on solving equations

Select one or two activities that were important in your understanding of how to solve equations.

• Other key activities

Include one activity on measuring sun shadows or trigonometry and one or two other activities that you think were important in developing the key ideas of this unit.

continued ▶

- A Problem of the Week

  Select one of the four POWs you completed during this unit (*Cutting the Pie, Pool Pockets, Trying Triangles,* and *Spiralaterals*).

- Other quality work

  Select one or two other pieces of work that demonstrate your best efforts. (These can be any work from the unit, such as a Problem of the Week, an individual or group activity, or a presentation.)

End-of-Year Review

Because this is the final unit, take this opportunity to look back at your overall experience. Here are some questions you might want to answer.

- How was this experience different from your previous work in mathematics? Did you learn the mathematics differently? How was the mathematics itself different?

- How have you changed personally as a result of your experience? Has your confidence in your own ability grown? How has your experience of working with others changed?

- What are your mathematics goals for the rest of your high school years? How have those goals changed over the past year and why?

Include any other thoughts you might like to share with a reader of your portfolio.

SUPPLEMENTAL ACTIVITIES

Many of the supplemental activities for *Shadows* focus on similarity, but other activities look at logic and counterexamples, at the use of proportions, or at the history underlying concepts from the unit. These are some examples.

- *Fit Them Together* and *Similar Areas* examine the connection between area and similarity.

- *Is It Sufficient?* and *Triangular Data* look at the conditions needed for similarity and congruence. They also ask you to find counterexamples or generalizations.

- *Proportions Everywhere* uses proportionality in a geometric setting. *What If They Kept Running?* looks at proportionality in the context of a track competition.

- *The Parallel Postulate* gives you a glimpse into the historical development of ideas about parallelism. You'll read how these ideas led to the development of non-Euclidean geometry.

Some Other Shadows

The central problem in this unit involves finding the length of a shadow cast on the floor by a lamp or cast on the ground by a streetlight. But what if the shadow is cast along a surface that isn't horizontal? For example, the picture shows someone using a lamp to create shadows on a wall.

These other kinds of shadows may be difficult to study, but give it a try. You don't have to look at the length of the shadow—you can investigate any aspect you like.

Investigation

In each investigation, it was important that everything was the same in each experiment except for the variable you were testing. In other words, each experiment had one variable (such as the distance from the lamp or the weight of the bob) that you were changing, and one result (such as the length of the period or the length of the shadow) that you looked at to see if there was an effect.

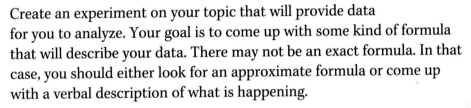

For this activity, your task is to think of your own topic of investigation. In your investigation, as before, there should be one variable that changes (the independent variable) and one result (the dependent variable) where you look for an effect.

Create an experiment on your topic that will provide data for you to analyze. Your goal is to come up with some kind of formula that will describe your data. There may not be an exact formula. In that case, you should either look for an approximate formula or come up with a verbal description of what is happening.

Here is a suggested plan.

1. Decide on a topic to investigate. Pick a topic that can be organized so that there is one independent variable and one dependent variable. Everything else should remain fixed.

2. Design the experiment.

3. Do one or two trial runs of your experiment. Refine it as necessary.

4. Collect and organize the data.

5. Try to use your data to come up with a formula or verbal description of what is happening.

You will need to turn in a summary of your work. You can use these steps as an outline.

Cutting Through the Layers

Imagine a single piece of string that can be folded back and forth. In the first picture, the string is folded so that it has three "layers." But it is still one piece of string.

Now imagine that you take scissors and cut across the folded string, as indicated by the dotted line. The result will be four separate pieces of string.

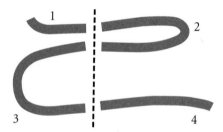

You could have made more than one cut across the folded string, creating more pieces.

In the third picture, two cuts have been made, creating a total of seven pieces.

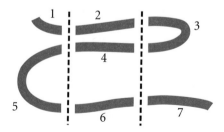

continued ▶

You could also start with more layers in the folded string. Four layers and three cuts create a total of 13 pieces.

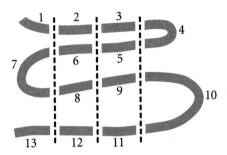

All of this information about layers, cuts, and pieces of string has been organized into an In-Out table. Because the number of pieces depends on both the number of layers and the number of cuts, there are two inputs, while the number of pieces is the output.

Inputs		Output
Number of layers	Number of cuts	Number of pieces
3	0	1
3	1	4
3	2	7
4	3	13

1. Make your own pictures of string with different numbers of layers and different numbers of cuts. Count the pieces and add that information to the In-Out table.

2. Suppose the number of layers is L and the number of cuts is C. Find a rule or formula expressing the number of pieces as a function of both L and C. In other words, tell what to do with L and C to find out how many pieces there will be.

Explaining the Layers

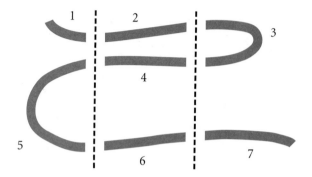

In *Cutting Through the Layers,* you looked at diagrams of a single piece of string folded back and forth into several "layers." A number of cuts were made across the folded string. The problem was to figure out how many pieces of string you get. For example, the diagram shows three layers of string and two cuts, resulting in seven pieces of string after the cuts.

Your task in that activity was to find a formula for the number of pieces of string in terms of the number of layers (represented by L) and the number of cuts (represented by C).

Your task in this activity is to write an explanation, in terms of the problem situation, for why your formula works for any number of layers and any number of cuts.

Crates

In *An N-by-N Window* and in *More About Windows,* you looked for formulas for the amount of wood strip needed to create a window frame. Now suppose you are building a frame for a wooden crate, such as this one.

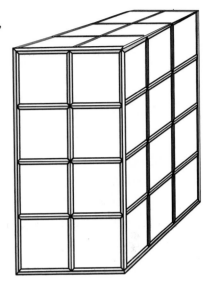

The wood strip is used only to make a frame for the crate. Other wood panels, each 1-foot square, have been placed between the strips to make a solid box.

1. Find the amount of wood strip needed for a crate 2 feet wide, 3 feet long, and 4 feet high.

   Only three sides of the crate are shown, so you'll need to imagine the other three sides. Remember that some of the wood strips are shared by two sides of the crate.

2. Look for a general formula for a crate *w* feet wide, *l* feet long, and *h* feet high.

Instruct the Pro

On a clean sheet of paper make a sketch of a figure made up only of line segments that touch end to end. Your sketch can be more interesting than this one.

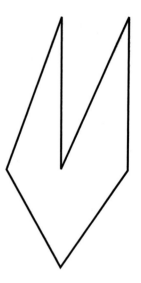

Write instructions so that someone with a ruler and a protractor could draw a diagram exactly like yours without seeing your diagram.

Select a point in your diagram where two segments meet. Use that as the starting point for your instructions. You may also want to tell the person where to start on the sheet of paper. For example, you might say, "Start in the center of your page."

Scale It!

Scale drawings and scale models are used for many purposes, such as map-making and testing new inventions.

Here are a few options you might consider for an investigation of scaling.

- Make a scale drawing of some aspect of your school or neighborhood.
- Build a scale model of some object.
- Interview someone who uses scale drawings or scale models in his or her work.

If you have other ideas for learning more about scaling, that's fine too. Whatever you do, write a report about your work. In your report, discuss how the ideas of this unit were connected to your investigation.

The Golden Ratio

There is a ratio that often comes up in art and nature called the *golden ratio* or the *golden section*. Many people think structures that make use of this ratio are visually appealing.

Research the golden ratio in nature and in things people create. Write at least two pages on what it is and where it occurs. Include a list of the sources of your information.

From Top to Bottom

A **scale drawing** of a figure or object is a drawing that is geometrically similar to the original. In this activity, your task is to make two scale drawings of pentagon *ABCDE*. In each drawing, keep the pentagon facing the same way as shown here.

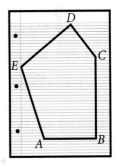

Make your two drawings on $8\frac{1}{2}$-inch-by-11-inch paper, with one drawing on each side of the sheet. Use a ruler and a protractor to carefully measure the sides and angles. Make your drawings as accurate as you can.

1. Make a scale drawing of the pentagon so that the bottom of your drawing (corresponding to side *AB*) is 1 inch from the bottom of the page and the top of your drawing is 1 inch from the top. Your drawing should be positioned roughly like this.

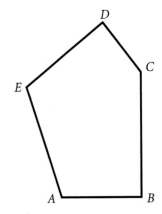

2. For your second drawing, flip over the sheet of paper and turn it sideways. With the paper in this position, make another scale drawing of the pentagon. Again, the bottom of the pentagon should be 1 inch from the bottom of the page and the top of the pentagon should be 1 inch from the top. The pentagon should be positioned roughly like this.

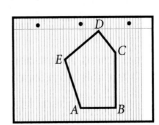

3. For each drawing, find the ratio between the length of a side of your drawing and the length of the corresponding side of the pentagon in this book.

   a. What is the ratio for the first drawing?

   b. What is the ratio for the second drawing?

   c. How might you find these ratios before you start the drawings?

Proportions Everywhere

Polygons *ABCDE* and *PQRST* are similar, so the ratios of the lengths of corresponding sides are the same. For example, the ratio $\frac{AB}{PQ}$ is the same as the ratio $\frac{CD}{RS}$, because $\overline{AB}$ corresponds to $\overline{PQ}$ and $\overline{CD}$ corresponds to $\overline{RS}$.

What about other lengths in these diagrams? Does every part of the first figure have a corresponding part in the second? Do other ratios of corresponding parts come out the same as these ratios? First look at some examples.

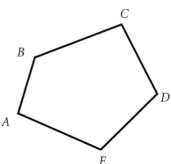

1. Consider the diagonal $\overline{BE}$ in the first polygon.

    a. What segment in the second polygon corresponds to this diagonal?

    b. Is the ratio of the length of $\overline{BE}$ to the length of its corresponding part the same as the ratio of *AB* to *PQ*? Why or why not?

2. Use *F* to represent the midpoint of side $\overline{AE}$.

    a. What segment in the second polygon corresponds to $\overline{CF}$?

    b. Is the ratio of the length of $\overline{CF}$ to the length of its corresponding part the same as the ratio of *AB* to *PQ*? Why or why not?

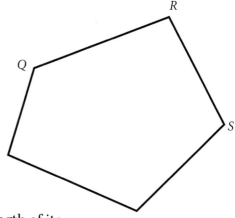

Now make up your own examples and think about generalizing. Clearly state any general principles you think must hold true.

How Can They Not Be Similar?

You know that two polygons are similar if they satisfy both of these conditions.

- The angles of the first polygon are equal to the corresponding angles of the second.
- The sides of the first polygon are proportional to the corresponding sides of the second.

What if the word *corresponding* is omitted? Do the polygons still have to be similar?

The answer is no, but constructing a counterexample is not easy. To make life simpler for you, look at the special case where the ratio of side lengths is 1. In other words, your task is to try to construct two polygons that satisfy both of the following conditions but that are not similar.

- The angles of the first are equal (as a group) to the angles of the second.
- The lengths of the sides of the first polygon are equal (as a group) to the lengths of the sides of the second.

Rigidity Can Be Good

You saw in *Why Are Triangles Special?* that if you build a triangle using specific lengths for the sides, then the triangle is rigid. That is, once the sides are put together, there is no flexibility.

For polygons with more than three sides, the polygon can be "flexed" at the corners. This changes the shape without changing the lengths of the sides. Of course, in many situations, it's important to be flexible. But in architecture, it can be important that a building keep its shape.

Investigate the significance of rigidity for architecture and construction. You may want to do some construction of your own, or you may want to read about how triangles can be used to make buildings stable.

Look up the work of R. Buckminster Fuller.

Is It Sufficient?

As you already know, you can conclude that two triangles are similar whenever you know that two angles of one triangle are equal to two angles of the other triangle.

Mathematicians express this by saying that having two pairs of equal angles is *sufficient* for concluding that two triangles are similar.

Your goal in this activity is to investigate what other information about two triangles can be considered sufficient to conclude that the triangles are similar.

For each of the following conditions, start out as a skeptic. Use your ruler and protractor to make drawings, and try to find two triangles that fit the condition but that are not similar. In other words, look for a counterexample. If you find such triangles, you will have shown that the particular combination is not sufficient to conclude that the triangles are similar.

On the other hand, you may decide that a given condition is sufficient—that is, that there are no counterexamples. In that case, try to explain why any two triangles that fit the condition must be similar.

continued ▶

Condition 1

An angle of one triangle is equal to an angle of the other triangle.

Condition 2

A side of one triangle is proportional to a side of the other triangle.

Condition 3

A pair of sides of one triangle is proportional to a pair of sides of the other triangle.

Condition 4

The three sides of one triangle are proportional to the three sides of the other triangle.

Condition 5

An angle of one triangle is equal to an angle of the other triangle, and a side of one triangle is proportional to a side of the other triangle.

Condition 6

A pair of sides of one triangle is proportional to a pair of sides of the other triangle, and the angles between these pairs of sides are equal.

Condition 7

A pair of sides of one triangle is proportional to a pair of sides of the other triangle, and an angle not between the pair in one triangle is equal to the corresponding angle of the other triangle.

Triangular Data

You saw in *Why Are Triangles Special?* that if you are given three lengths, there is at most one way to build a triangle with sides of those lengths. Mathematicians express this property by saying that the lengths of the sides *determine* the triangle.

Here's another way to express this property.

> If the sides of one triangle have the same lengths as the sides of another triangle, then the two triangles must be congruent.

We sometimes think of a triangle as having six parts—three sides and three angles. Three sides determine a triangle. You saw in *Are Angles Enough?* that three angles do not determine a triangle, because two triangles with the same angles are similar but may not be congruent.

In Part I of this activity, you will explore what other combinations of information determine a triangle. In Part II, you will try to generalize your discoveries.

Part I: Exploring Triangles

In each problem, you are given the values for three of the six parts of a possible triangle, $\triangle ABC$. Your task is to try to draw a triangle that fits the conditions. In each case, answer two questions.

a. Is it possible to draw a triangle that fits the conditions?

b. If so, will two triangles that both fit the conditions have to be congruent?

In other words, your task is to find out if the given information determines a triangle.

Answer questions a and b for problems 1 through 8. Justify your answers.

continued ▶

Note: You will probably find it helpful to use a ruler and a protractor.

1. $AC = 5$ inches, $\angle ABC = 50°$, and $\angle CAB = 110°$
2. $BC = 7$ inches, $\angle ABC = 70°$, and $\angle BCA = 80°$
3. $AB = 5$ inches, $BC = 8$ inches, and $\angle ABC = 70°$
4. $AB = 7$ inches, $AC = 11$ inches, and $\angle CAB = 130°$
5. $AB = 7$ inches, $AC = 6$ inches, and $\angle ABC = 50°$
6. $AB = 2$ inches, $BC = 4$ inches, and $\angle BCA = 60°$
7. $AB = 3$ inches, $AC = 5$ inches, and $\angle ABC = 90°$
8. $AB = 4$ inches, $BC = 6$ inches, and $\angle CAB = 100°$

Part II: Generalizing Triangles

Look at the preceding examples and consider these questions.

- How many lengths and how many angles were provided in each case?
- How were the lengths and angles situated in relation to each other in the triangle?

Based on these questions and others you might ask on your own, try to develop some general principles. The central question to explore is, What type of information determines a triangle?

What If They Kept Running?

Mandy, Ivy, and Charley were running a 1500-meter race. Mandy won.

At the moment when Mandy crossed the finish line, she was 300 meters ahead of Ivy and 420 meters ahead of Charley.

If Ivy and Charley keep running at the same rates as they have run so far, how many meters ahead of Charley will Ivy be when Ivy crosses the finish line?

Adapted from *Mathematics Teacher* (January 1990), National Council of Teachers of Mathematics, volume 83, number 1.

An Inside Proof

You have seen that you can create similar triangles by drawing segments within a triangle. In this activity, your task is to prove some conclusions that you may have seen in *More Similar Triangles.*

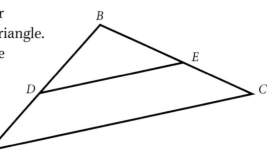

In this diagram, *ABC* is an arbitrary triangle, *D* is a point on $\overline{AB}$, and *E* is a point on $\overline{CB}$.

Suppose that $\overline{DE}$ is parallel to $\overline{AC}$.

1. Prove that $\triangle DBE$ is similar to $\triangle ABC$.

2. Prove that if *D* is the midpoint of $\overline{AB}$, then *E* is the midpoint of $\overline{BC}$.

Suggestion: Use principles from *More About Angles.*

Fit Them Together

The diagram shows how to start with one triangle (the top triangle) and fit four copies of it together to make a larger version of that triangle.

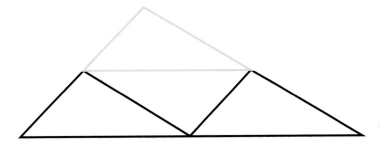

The diagram shows that the large triangle is similar to the original triangle, because each side is twice as long as its corresponding side of the original. The diagram also shows that the area of the "double-size" triangle is four times the area of the original triangle.

1. Can you do this "fitting together" starting with any triangle? Try to find a triangle for which you can't put four copies together this way. Or explain why this diagram works for any initial triangle.

2. Move on to quadrilaterals. If you start with any quadrilateral, can you fit four exact copies of it together to make a double-size version of that quadrilateral? The answer may depend on the quadrilateral.

   Start with squares, and then try rectangles, parallelograms, **trapezoids,** and others. In each case, explore whether four copies can be put together to make a double-size version.

   If you think that this can be done with a given category of quadrilaterals, show how. If you find quadrilaterals for which this fitting together cannot be done, show some of them.

3. What about area? If you start with any quadrilateral and make a similar quadrilateral with sides twice as long as those of the original, will the area of this larger quadrilateral always be four times the area of the original? Either explain why this is so, or give a counterexample.

Similar Areas

In *Fit Them Together,* you saw that you could double the lengths of the sides of a triangle to create a new triangle that was similar to the original and that had an area four times that of the original.

What more can you say about areas of similar figures?

Specifically, how does the ratio of the area of two similar figures depend on the ratio of the lengths of corresponding sides?

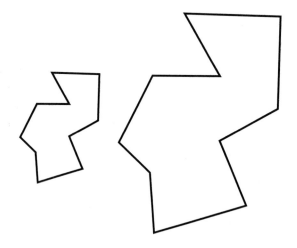

The Parallel Postulate

When the Greek mathematician Euclid wrote his *Elements* (in about 300 BCE), he made certain fundamental geometric assumptions, called *postulates* and *axioms*. His first postulate says:

> A straight line may be drawn between any two points.

That's pretty reasonable. Most of his postulates are just as simple as this first one. But Euclid's fifth postulate is quite complicated.

> If two straight lines lying in a plane are met by another line, and if the sum of the interior angles on one side is less than two right angles, then the straight lines, if extended sufficiently, will meet on the side on which the sum of the angles is less than two right angles.

1. What does Euclid's fifth postulate mean? Draw a diagram to explain what he was talking about. The phrase "another line" corresponds to what was called a *transversal* in the activity *More About Angles*. In other words, this line crosses each of the other two lines.

2. Explain how this postulate is related to the principles about parallel lines that you found in *More About Angles*. In particular, discuss the connection between Euclid's fifth postulate and the statement, "Given a line *L* and a point *P* not on *L,* there is a unique line through *P* that is parallel to *L*."

   This statement is sometimes called *Playfair's axiom,* after the Scottish mathematician John Playfair (1748–1819).

Mathematicians were intrigued by Euclid's fifth postulate, partly because it was so complicated. A long history developed in which they tried to prove this statement without making any assumptions except for Euclid's other postulates. They were never successful.

continued ▶

In the nineteenth century, three mathematicians in three different countries showed this challenge was impossible. A German mathematician, Carl Friedrich Gauss (1777–1855), a Russian mathematician, Nicolai Ivanovitch Lobachevsky (1793–1856), and a Hungarian mathematician, Janós Bolyai (1802–1860), all explored the idea of a geometric system in which Euclid's fifth postulate was false. Independently of one another, they discovered that such a system was logically possible. The system they developed is called *hyperbolic geometry*.

Another geometric system, which required changing some of Euclid's other postulates, was developed by the German mathematician Georg Friedrich Bernhard Riemann (1826–1866). This system is called *elliptic geometry*.

Both hyperbolic and elliptic geometry are examples of geometric systems that describe the geometry of surfaces other than a plane. Both are also examples of a broader field called *non-Euclidean geometry*. It turns out that Albert Einstein made use of the ideas of non-Euclidean geometry in developing his concepts of space and time that are part of the theory of relativity.

3. Read about and report further on the history of Euclid's fifth postulate and the development of non-Euclidean geometry.

Exterior Angles and Polygon Angle Sums

In *A Parallel Proof,* you saw how to use principles about parallel lines to prove that the sum of the angles of any triangle is exactly 180°. You also know that this fact for triangles can be used to develop an angle-sum formula for arbitrary polygons. It turns out that these facts about angle sums can also be proved by using a concept called **exterior angles.**

This activity will explain that concept and then ask questions to help you develop the proof.

Exterior Angles

In general, when we talk about the angles of a polygon, we mean the angles inside the figure. For example, in the first pentagon, we might mean ∠ABC or ∠DEA, among others.

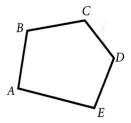

You get some other interesting angles by extending the sides of the polygon.

The second diagram shows the same polygon with $\overline{BC}$ extended beyond *C* to point *F*.

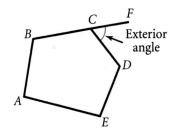

Angle *FCD* is called an exterior angle of the polygon. We sometimes call an angle inside a polygon, such as ∠BCD, an **interior angle.**

There are actually two exterior angles at each vertex of a polygon. For example, at *C*, in addition to extending $\overline{BC}$, you could also extend $\overline{DC}$ past *C*. Both ∠BCK and ∠DCF are exterior angles at *C* for polygon *ABCDE*.

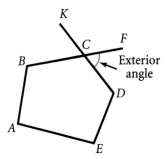

In this activity, you will consider only one exterior angle at each vertex. Because angles *BCK* and *DCF* are **vertical angles,** they are equal in size. Therefore, it won't matter which one you use.

continued ▶

Polygon Angle Sums

The proof of the angle-sum formula for polygons combines two ideas.

- The relationship between each interior angle and the corresponding exterior angle
- A general formula about exterior angles

You will develop the two ideas separately and then put them together.

Step 1. Interior and Exterior Angles

Step 1, the easier of the two parts, has two stages. First answer this question.

> *What is the relationship between an interior angle of a polygon and the corresponding exterior angle? For example, how are angles BCD and FCD related?*

Write a formula or equation relating two such angles. Explain why that relationship must hold true for every such pair of angles.

Your answer to the preceding question should give you a formula for the sum of each interior angle and the corresponding exterior angle. Basing your response on that result, answer this question.

> *If you combine all the interior angles with their corresponding exterior angles, using only one exterior angle for each vertex, what is the sum of all of these angles?*

The answer depends on the number of sides of the polygon.

Step 2. Total Turns

The second key idea involves only the exterior angles of a polygon, choosing one at each vertex, as shown in the diagram.

Imagine that you are standing at point *C*, facing toward *F*. From that position, turn toward *D*, turning through an angle equal to the exterior angle *FCD*.

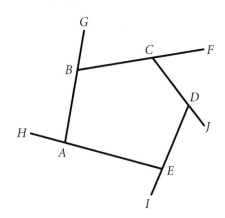

continued ▶

Move along $\overline{CD}$ to point D. When you get there, turn toward E, this time turning through an angle equal to exterior angle JDE.

Move along $\overline{DE}$ to E, and again make a turn. Continue around the polygon in this way, moving along each side and then turning through the exterior angle.

Your final turn will occur when you are at point B, facing toward G. You will turn toward C, going through a turn equal to the last of the exterior angles. You are now facing in the same direction in which you started.

Answer the key question here.

What is the total number of degrees in all of your turns? In other words, what is the sum of all the exterior angles of the polygon (taking one exterior angle at each vertex)?

Would your result work for any polygon? Explain.

Step 3. Putting It All Together

In Step 1, you found a formula for the sum of an interior angle and its corresponding exterior angle. From that you should have found a formula for the sum of all the interior and exterior angles for a polygon with any number of sides.

In Step 2, you found a formula for the sum of all the exterior angles.

The final step of the process is this.

What is the sum of the interior angles of the polygon?

Exactly One-Half!

You may have observed that according to your calculator,
sin 30° = 0.5000.

You might wonder whether this is just an approximation or if sin 30° is really exactly one-half. Your task in this activity is to show, beyond any doubt, that the sine of 30° is exactly, precisely one-half.

Use an equilateral triangle and the fact that the sum of the angles of a triangle is 180°.

Eye Exam and Lookout Point

Eye Exam

An eye surgeon must perform an operation on a person who has pressure behind the cornea. (The cornea is the shaded area in the picture.) The surgeon will use a laser to make small holes along the edge of the cornea.

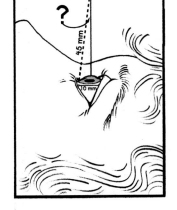

The patient will be lying on the operating table, and the laser will be above her. More precisely, the laser will be directly above her pupil (the center of her eye). The laser is 45 millimeters from the outer edge of the cornea. The diameter of the cornea is 10 millimeters.

The surgeon needs to find the angle at which to set the laser. This is the angle shown in the diagram as "?".

1. Use trigonometry to set up an equation that could be used to find this angle.

2. Use your calculator to get an approximate solution to this equation.

Lookout Point

The Park Service has created a lookout point over a beautiful valley that contains some interesting natural features. The service will be posting signs telling visitors how to adjust the lookout point telescope to find these features.

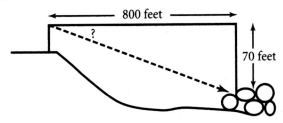

One very interesting rock formation is 70 feet lower than the telescope and 800 feet out from the telescope.

The telescope always resets to horizontal before each visitor uses it. By how big an angle should the Park Service tell people to tilt the telescope in order to see this rock formation?

Pole Cat

Poor Diane. Her cat, Wanda Ann, has climbed up a telephone pole and can't get down.

The crossbar of the telephone pole is 20 feet high. Diane is 5 feet 6 inches tall and can reach 1 foot above her head. She has a 15-foot ladder. To keep the ladder from tipping, Diane must lean it against the pole at an angle of 70° to the ground. Can Diane save Wanda Ann?

meow...

Dog in a Ditch

Oscar and Rasheed are identical twins. They are both 6 feet tall. They are on opposite sides of a ditch that is 30 feet wide. Their dog, Earl, is at the bottom of the ditch.

Earl can get out if he wants to. When Oscar looks at Earl, his line of sight makes a 75° angle below the horizontal. When Rasheed looks at Earl, his line of sight makes a 40° angle below horizontal. How deep is the ditch?

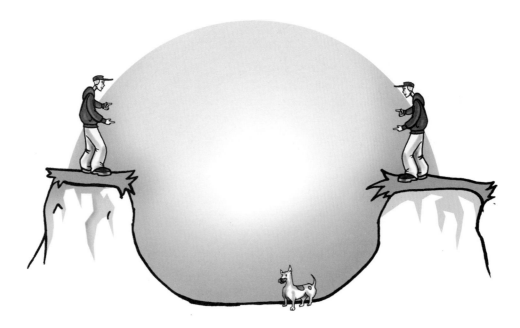

GLOSSARY

This is the glossary for all five units of IMP Year 1. This glossary may be useful when you encounter a term in **bold** text that is new or unfamiliar, or it can be used to confirm or clarify your understanding of a term.

Absolute value   The distance a number is from 0 on the number line. The symbol | | stands for the absolute value of that number.

   Examples: $|-2| = 2; |7| = 7; |0| = 0$

Acute angle   An angle that measures more than 0° and less than 90°.

Acute triangle   A triangle with angles that are all acute.

Adjacent angles   Two angles with the same vertex, formed using a shared ray.

   Example: Angles A and B are adjacent angles.

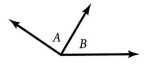

Adjacent side (for an acute angle of a right triangle)   The side of the **right triangle** that together with the hypotenuse forms the given angle.

   Example: In the right triangle ABC, side $\overline{BC}$ is adjacent to $\angle C$, and side $\overline{AB}$ is adjacent to $\angle A$.

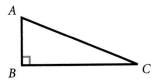

Alternate interior angles   If two lines are intersected by a **transversal,** then the inside angles with different vertices that are on opposite sides of the transversal are alternate interior angles.

> *Example:* Angles *K* and *L* are one pair of alternate interior angles, and angles *M* and *N* are another pair.

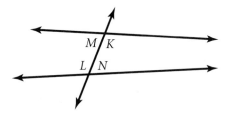

Amplitude (for a pendulum)   The angle of a pendulum's swing, measured from the vertical to the most outward position of the pendulum during its swing.

> *Example:* The pendulum in the diagram has an amplitude of 20°.

Angle   Informally, an amount of turn, usually measured in **degrees.** Formally, the geometric figure formed by two **rays** with a common point, called the **vertex** of the angle.

Angle of elevation   The angle at which an object appears above the horizontal, as measured from a chosen point.

> *Example:* The diagram shows the angle of elevation to the top of the tree from point *A*.

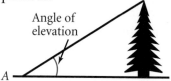

Area   Informally, the amount of space inside a two-dimensional figure, usually measured in square units.

Area model   For probability, a diagram showing the possible outcomes of a particular event. Each portion of the model represents an outcome, and the ratio of the area of that portion to the area of the whole model is the probability of that outcome.

Average   Often refers to the **mean,** but more generally refers to some sort of measure of the center of a data set.

Axis (plural: axes)   See **coordinate system.**

Box-and-whiskers plot (or box plot)   A display that shows the minimum, **lower quartile, median, upper quartile,** and maximum of a data set.

Closed formula   A method to figure the *Out* value of a table by using its associated *In* value. Contrast with **recursive formula.**

Coefficient   Usually, a number used to multiply a variable or power of a variable in an algebraic expression.

   Example: In the expression $3x + 4x^2$, 3 and 4 are coefficients.

Complementary angles   A pair of angles with measures that add to 90°. If two complementary angles are adjacent, together they form a right angle.

Conclusion   Informally, any statement arrived at by reasoning or through examples. See also **"If . . . , then . . ." statement.**

Conditional probability   The probability that an event will occur based on the assumption that some other event has already occurred.

Congruent   Informally, having the same shape and size. Formally, two polygons are congruent if their corresponding angles have equal measure and their corresponding sides have equal length. The symbol $\cong$ means "is congruent to."

Conjecture   A theory or an idea about how something works, usually based on examples.

Consecutive sum   A way to write a number as the sum of a sequence of two or more **whole numbers,** where each number of the sequence is one more than the previous number.

   Example: $4 + 5 + 6$ is a consecutive sum. This consecutive sum shows a way to represent the number 15.

Constraint   Informally, a limitation or restriction.

Continuous graph   Informally, a graph that can be drawn without lifting the pencil, in contrast to a **discrete graph.**

Coordinate system   A way to represent points in the plane with ordered pairs of numbers called coordinates. The system is based on two perpendicular number lines, one **horizontal** and one **vertical,** called coordinate axes. The point where the lines intersect is called the **origin.** Traditionally, the axes are labeled with the variables x and y, as shown below. The horizontal axis is often called the x-axis, and the vertical axis is often called the y-axis.

Example: Point A has coordinates $(3, -2)$.

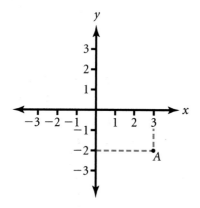

Corresponding angles (for a transversal)   If two lines are intersected by a **transversal,** then two angles are corresponding angles if they occupy the same position relative to the transversal and the other lines that form them.

Example: Angles A and D are a pair of corresponding angles, and angles B and E are another pair of corresponding angles.

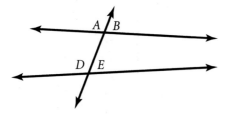

Corresponding parts (for similar or congruent polygons)   Sides or angles of polygons that have the same relative position.

Example: Side a in the small triangle and side b in the large triangle are corresponding parts.

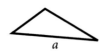
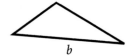

Cosine   The ratio of the length of the leg **adjacent** to one non-right angle of a right triangle to the length of the **hypotenuse.** The cosine of $\angle A$ is abbreviated as cos A.

$$\cos A = \frac{\text{length of leg adjacent to } \angle A}{\text{length of hypotenuse}}$$

Counterexample   An example that demonstrates a conjecture is not true.

Counting number   See **natural number.**

Degree   The measurement unit for an angle defined by having a complete turn equal to 360 degrees. The symbol ° represents degrees.

Dependent variable   Informally, the changing quantities represented by the *Out* values in an In-Out table. Also the quantity that changes as a result of changes in the **independent variable.** On a graph, the name of the dependent variable is labeled on the *y*-axis.

Diagonal   In a polygon, a line segment that connects two vertices and is not a side of the polygon.

Discrete graph   A graph consisting of isolated or unconnected points, in contrast to a **continuous graph.**

Divisor   A factor of an integer.

   Example: 1, 2, 3, 4, 6, and 12 are the positive divisors of 12.

Domain   The set of values that can be used as inputs for a given function.

Edge   The line segment where two faces of a three-dimensional shape intersect.

   Example: When a die is rolled, it never lands teetering on one of its edges.

Equation   A statement, using an equal sign, of two **equivalent expressions.** See also **formula, function.**

Equilateral triangle   A triangle with all sides the same length.

Equivalent equations Two or more equations that have the same solution (or solutions).

Example: The equation $3x + 5 = 17$ is equivalent to the equations $3x = 12$ and $x = 4$. When the solution, 4, is substituted for x, each equation is true ($17 = 17$, $12 = 12$, or $4 = 4$).

Example: The equations $y = 5 + 2x$ and $2y = 4x + 10$ are equivalent equations. Any coordinate pair (x, y) that makes the first equation true will also be true in the second equation, such as $(1, 7)$, $(2, 9)$, or $(3, 11)$,

Equivalent expressions Two or more expressions that have the same value when any number is **substituted** for the variable.

Example: The expression $4 + 6x$ is equivalent to $2(2 + 3x)$. When any number (for example, 5) is substituted for x, both expressions evaluate to the same number (in the example, 34).

Event The specific set of outcomes from performing an experiment several times, like ways a pair of dice could show a total value of 5. See also **outcome.**

Expected value In a game or other probability situation, the average amount gained or lost per turn in the long run.

Experimental probability See **observed probability.**

Exponent A number written as a **superscript** to another number (the base), to indicate how many times the base is used as a factor of multiplication.

Example: $2^5 = 2 \cdot 2 \cdot 2 \cdot 2 \cdot 2$. The 5 is the exponent, with 2 as its base. 2 is used as a factor 5 times.

Expression The written combination of variables and numbers that often represents some situation.

Exterior angle An angle formed outside a polygon by extending one of its sides.

Example: The diagram shows $\angle BAX$, an exterior angle for polygon $ABCDE$.

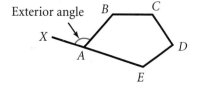

Face   A flat surface of a three-dimensional shape.

Example: The face of a die is where the numbers are printed.

Factorial   The product of all the whole numbers from a particular number down to 1. The symbol ! stands for factorial.

Example: 5! (read "five factorial") means $5 \cdot 4 \cdot 3 \cdot 2 \cdot 1$.

Fair game   A game in which both players are expected to come out equally well in the long run.

Formula   A mathematical statement describing a relationship among variables or indicating how to calculate for some unknown. See also **equation, function.**

Frequency bar graph   A bar graph showing how often each result occurs.

Example: This frequency bar graph shows, for instance, that 11 times in 80 rolls, the sum of two dice was 6.

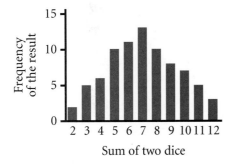

Function   Informally, a process or rule for determining the numerical value of one variable in terms of another. A function is often represented as a set of number pairs in which the second number is determined by the first, according to the function rule.

Graph   A mathematical diagram for displaying information.

Hexagon   A polygon with six sides.

Horizontal   Extending side to side, like the horizon.

Hypotenuse   The longest side in a right triangle, or the length of this side. The hypotenuse is located opposite the right angle.

Example: In right triangle *ABC*, the hypotenuse is $\overline{AC}$.

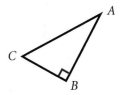

Hypothesis   Informally, a theory about a situation or about how a certain set of data is behaving. Also, a set of assumptions used to analyze or understand a situation. See also **"If . . . , then . . ." statement.**

"If . . . , then . . ." statement   A specific form of mathematical statement, saying that if one condition is true, then another condition must also be true.

>*Example:* If two angles of a triangle have equal measure, then the sides opposite these angles have equal length.
>
>The condition "two angles of a triangle have equal measure" is the **hypothesis.** The condition "the sides opposite these angles have equal length" is the **conclusion.**

Independent events   Two (or more) **events** are independent if the result of one does not influence the result of the other.

Independent variable   Informally, the changing quantities represented by the *In* values in an In-Out table. Also the quantity that, when changed, causes changes in the *Out* values, or the **dependent variable.** On a graph, the name of the independent variable is labeled on the *x*-axis.

>*Example:* In a situation involving time and distance traveled, time is usually the independent variable.

Integer   Any number that is either a counting number, 0, or the opposite of a counting number. The integers can be represented using set notation as

$$\{ \cdots -3, -2, -1, 0, 1, 2, 3, \cdots \}$$

>*Examples:* $-4, 0,$ and 10 are integers.

Interior angle   An angle inside a figure, especially within a polygon, formed by sides of the figure.

>*Example:* Angle *BAE* is an interior angle of the polygon *ABCDE.*

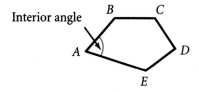

Isosceles triangle   A triangle with at least two sides of equal length.

Justify To make an argument for; to give a reason or explanation.

Leg Either of the two shorter sides in a right triangle. The two legs of a right triangle form the right angle of the triangle.

Linear function A **function** whose graph is a line. A common form for writing a linear function is $f(x) = ax + b$, where a is the **rate of change** and b is the y-coordinate of the **starting point.**

Line of best fit Informally, the line that comes closest to fitting a given set of points on a **discrete graph.**

Line segment The portion of a straight line between two given points.

Lower quartile The median of the values below the median of a data set.

Mathematical model A mathematical description or structure used to represent how a real-life situation works.

Mean The numerical average of a data set, found by adding the data items and dividing by the number of items in the set.

> *Example:* For the data set 8, 12, 12, 13, and 17, the sum of the data items is 62. There are 5 items in the data set, so the mean is $62 \div 5$, or 12.4.

Measurement variation The situation of taking several measurements of the same thing and getting different results.

Median (of a set of data) The "middle number" in a set of data that has been arranged from smallest to largest. If the data set has an even number of values, the median is the mean of the two "middle numbers."

> *Example:* For the data set 4, 17, 22, 56, and 100, the median is 22, because it is the number in the middle of the ordered list.

Mode (of a set of data) The number that occurs most often in a set of data. A data set may have more than one mode.

> *Example:* For the data set 3, 4, 7, 16, 18, 18, and 23, the mode is 18.

Natural number Any of the numbers used for counting, such as 1, 2, 3, 4, and so on, but not including zero. Also called **counting number.**

Normal distribution   A certain, precisely defined set of probabilities that can often be used to approximate real-life events. Sometimes used to refer to any data set whose graph is approximately "bell-shaped."

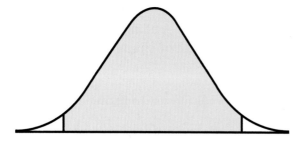

Observed probability   The likelihood of a certain event happening based on observed results, as distinct from **theoretical probability.** Also called **experimental probability.**

Obtuse angle   An angle that measures more than 90° and less than 180°.

Obtuse triangle   A triangle with an obtuse angle.

Octagon   An eight-sided polygon.

Opposite side   The side of a triangle across from a given angle.

Ordered pair   Two numbers paired together using the format (x, y), often used to locate a point in the **coordinate system.**

Order of operations   A set of conventions that mathematicians have agreed to use whenever a calculation involves more than one operation.

> *Example:* $2 + 3 \cdot 4$ is 14, not 20, because according to the conventions for order of operations, multiplication occurs before addition.

Origin   See **coordinate system.**

Outcome   The result of performing one trial of an experiment, like flipping a coin or drawing a card from a deck. See also **event.**

Parallel lines   Two lines in a plane that do not intersect.

Pentagon   A five-sided polygon.

Period   The length of time for a cyclical event to complete one full cycle.

Perpendicular lines   A pair of lines that intersect to form right angles.

Polygon   A closed, two-dimensional shape formed by three or more line segments. The line segments that form a polygon are called its **sides**. The endpoints of these segments are called **vertices** (singular: **vertex**).

Examples: All of these figures are polygons.

Prime number   A whole number greater than 1 that has only two whole-number divisors, 1 and itself.

Example: 7 is a prime number, because its only whole-number divisors are 1 and 7.

Probability   The likelihood of a certain event happening. For a situation involving equally likely outcomes, the probability that the outcome of an event will be an outcome within a given set is defined by a ratio:

$$\text{Probability} = \frac{\text{number of outcomes in the set}}{\text{total number of possible outcomes}}$$

Example: If a cube has 2 red faces and 4 green faces, the probability of rolling the cube and getting a green face is

$$\frac{\text{number of green faces}}{\text{total number of faces}} = \frac{4}{6}$$

Proof   An absolutely convincing argument.

Proportion   A statement that two ratios are equal.

Proportional   Having the same ratio.

Example: Corresponding sides of triangles *ABC* and *DEF* are proportional, because the ratios $\frac{4}{6}$, $\frac{8}{12}$, and $\frac{10}{15}$ are equal.

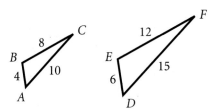

Quadrant   One of the four regions created in a coordinate system by using the *x*-axis and the *y*-axis as boundaries. The quadrants have standard numbering, as shown.

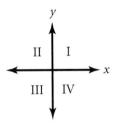

Quadrilateral   A four-sided polygon.

Random   Used in probability to indicate that any of several events is equally likely. In situations where events are not equally likely to occur, random is used to mean selection from a set of events according to a precisely described distribution.

Range (of a set of data)   The difference between the largest and smallest numbers in the set.

    Example: For the data set 7, 12, 18, 18, and 29, the range is 29 − 7, or 22.

Rate (or **rate of change**)   A number describing change. It is calculated by computing a **ratio** of two quantities.

    Examples: miles per day, heartbeats per minute.

Ratio   A comparison of two numbers.

Ray   The part of a line from a single point, called the **vertex**, through another point on the line and continuing infinitely in that direction.

Rectangle   A four-sided polygon with angles that are all right angles.

Recursive formula   A method to determine the *Out* value of a table by using the previous *Out* value in the table. This method requires the *In* values to be listed in order, going up by 1 at each step. Contrast with **closed formula.**

Regular polygon   A polygon with sides that all have equal length and angles that all have equal measure.

Rhombus   A four-sided polygon with sides that all have the same length.

Right angle   An angle that measures 90°.

Right triangle   A triangle with a **right angle.**

Sample standard deviation   The calculation on a set of data taken from a larger population of data, used to estimate the **standard deviation** of the larger population.

Sequence   An ordered list of numbers, expressions, or pictures, usually following a pattern or rule.

  Example: 1, 3, 5, 7, 9, . . . is the sequence of positive odd numbers.

Sigma notation   See **summation notation.**

Similar   Informally, having the same shape. Formally, two polygons are similar if their corresponding angles have equal measure and their corresponding sides are proportional in length. The symbol $\sim$ means "is similar to."

Simulation   An experiment or set of experiments using a model of an event that is based on the same probabilities as the real event. Simulations allow people to estimate the likelihood of an event when it is impractical to experiment with the real event.

Sine   The ratio of the length of the leg **opposite** one non-right angle of a right triangle to the length of the **hypotenuse.** The sine of $\angle A$ is abbreviated as sin A.

$$\sin A = \frac{\text{length of leg opposite } \angle A}{\text{length of hypotenuse}}$$

Slope   Informally, the steepness of a line.

Solution   A value that, when substituted for a variable in an equation, makes the equation a true statement.

  Example: The value $x = 3$ is a solution to the equation $2x = 6$ because $2 \cdot 3 = 6$.

Square   A four-sided polygon with all sides of equal length and with four right angles.

Square (of a number)   The number multiplied by itself; in other words, the number to the **exponent** 2.

Square root   A number whose square is a given number. The symbol $\sqrt{}$ is used to denote the nonnegative square root of a number.

> *Example:* Both 6 and -6 are square roots of 36, because $6^2 = 36$ and $(-6)^2 = 36$; $\sqrt{36} = 6$.

Standard deviation   A specific measurement of how spread out a set of data is, usually represented by the lowercase Greek letter sigma (σ).

Starting point   An informal reference to the amount of something at the beginning of a situation. The use of the word *point* emphasizes that this starting amount coincides with the *y*-coordinate associated with the **y-intercept.**

> *Example:* If a wagon train begins its westward journey with 40 pounds of salt, 40 is referred to as the starting amount, or starting point. The *y*-intercept of a graph of the salt usage would be (0, 40).

Stem-and-leaf plot (or stem plot)   A graphical display with "stems" showing the leftmost digit of the values separated from "leaves" showing the next digit or set of digits.

Straight angle   An angle that measures 180°. The rays forming a straight angle together make up a straight line.

Strategy   A complete plan about how to proceed in a game or problem situation. A strategy for a game should tell a person exactly what to do under any situation that can arise in the game.

Subscript   A symbol written below and to the right of another symbol.

> *Example:* For the variable P_B, the letter B is written as a subscript.

Substitute   To replace a variable in an **expression** with a numerical value. Usually followed by evaluation, that is, to compute the numerical value of the resulting expression.

Summary phrase   A concise phrase to describe the quantity represented by an expression.

Summation notation   A useful technique for writing the sum of a sequence of numbers. The uppercase Greek letter sigma (Σ) stands for summation.

> *Example:* The consecutive sum $3 + 4 + 5 + 6 + 7$ is $\displaystyle\sum_{r=3}^{7} r$.

Example: In the expression $\displaystyle\sum_{t=5}^{8}(4t^2 + 3)$, the number 5 is called the *lower limit,* the number 8 is called the *upper limit,* and the expression $4t^2 + 3$ is called the *summand.* The variable t is referred to as an *index variable,* or *dummy variable.*

Superscript   A symbol written above and to the right of another symbol, such as an **exponent.**

Supplementary angles   A pair of angles with measures that add to 180°. If two supplementary angles are adjacent, together they form a straight angle.

Tangent   The ratio of the length of the leg **opposite** one non-right angle of a right triangle to the length of the leg **adjacent** to the same angle. The tangent of $\angle A$ is abbreviated as tan A.

$$\tan A = \frac{\text{length of leg opposite } \angle A}{\text{length of leg adjacent to } \angle A}$$

Term (of an algebraic expression)   A part of an algebraic expression, combined with other terms using addition or subtraction.

Example: The expression $2x^2 + 3x - 12$ has three terms: $2x^2$, $3x$, and 12.

Term (of a sequence)   One of the items listed in a sequence.

Example: In the sequence 3, 5, 7, . . . , the number 3 is the first term, 5 is the second term, and so on.

Theoretical probability   The likelihood of an event occurring, as explained by a theory or model, as contrasted with **observed probability.**

Transversal   A line that intersects two or more other lines.

Example: The line ℓ is a transversal that intersects the lines m and n.

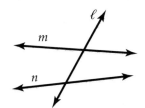

Trapezoid　A four-sided polygon with exactly one pair of parallel sides.

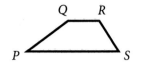

Example: Quadrilateral *PQRS* is a trapezoid, because $\overline{QR}$ and $\overline{PS}$ are parallel and $\overline{PQ}$ and $\overline{SR}$ are not parallel.

Triangle　A polygon with three sides.

Triangle inequality　The principle that the lengths of any two sides of a triangle must add up to more than the length of the third side.

Trigonometric function　Any of six functions defined for acute angles in terms of ratios of sides of a right triangle.

Unique　A word used to mean one and only one. Often used in reference to the only possible solution.

Upper quartile　The median of the values above the median of a data set.

Variable　A symbol used to represent a quantity that can have different values.

Variance　Like **standard deviation,** a specific measurement of how spread out a set of data is. The variance is the **square** of the **standard deviation.**

Vertex (plural: vertices)　A common endpoint of two segments or rays. See how this term is used in **angle, polygon,** and **ray.**

Vertical　Extending straight up and down, like a flagpole.

Vertical angles　A pair of opposite angles formed by a pair of intersecting lines.

Example: Angles *F* and *G* are vertical angles.

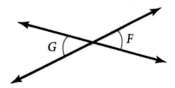

Whole number　A number that is either 0 or a **counting number.**

x-intercept　A place on a graph where a line or curve crosses the *x*-axis.

y-intercept　A place on a graph where a line or curve crosses the *y*-axis.

Index of Mathematical Ideas

The phrase "concept development" indicates that the associated term is previewed or developed on the page, but may not be named.

A

absolute value
    defined, 503
    concept development,
      205–206
acute angles, 34, 437, 503
    See also trigonometry
acute triangles, 503
addition
    with negative numbers, 22
    order of operations for, 19
    properties for. *See* real
      number properties
    See also consecutive sums;
      summation notation
additive identity, concept
    development, 21–23, 26, 56,
    57
adjacent, as term, 437–438, 503
algebraic expressions
    concept development, 30–31,
      43, 55, 69, 70–71, 72, 160,
      171, 186–187, 193, 194,
      195–196, 197, 198–199,
      200–201, 202–203, 246,
      268, 269–270, 271, 272,
      278, 285, 286, 291
    See also formulas; functions
algorithms
    with technology, 73, 126
    concept development, 13, 73,
      125, 126, 392–393, 451

alternate exterior angles,
    concept development, 440,
    443, 444
alternate interior angles
    defined, 504
    concept development, 440,
      443, 444, 445
amplitude (for a pendulum), 319,
    504
angle of approach, 447
angle of departure, 447
angle of elevation, 467, 504
angles
    acute, 34, 437, 503
    alternate interior. *See*
      alternate interior angles
    as "an amount of turn," 34
    of a polygon, 35, 495
    of approach, 447
    of a regular polygon, 38, 68
    corresponding. *See*
      corresponding angles;
      corresponding parts
    defined, 504
    of departure, 447
    of elevation, 467, 504
    exterior. *See* exterior angles
    interior. *See* interior angles
    measurement with
      calculators, 462
    measurement in degrees, 34,
      462

conjectures
    defined, 505
    process for exploring, 59
    concept development, 28–29, 36, 37,
        115, 134, 136, 161, 303
    See also proofs
consecutive, defined, 28
consecutive sums
    defined, 28, 505
    finding, 28–29, 33, 59, 62
    integers used in, 62
    summation notation for, 30–31
constraints
    defined, 505
    concept development, 180–182, 184, 185,
        186–187, 188, 197, 231–232, 274–275
continuous graphs
    as choice for representation, 210
    defined, 506
    concept development, 205–206, 210,
        211–212, 213
coordinate system
    defined, 506
    ordered pairs plotted on, 213, 512
    concept development, 205–206, 207,
        208–209, 210, 211–212, 213, 277
corresponding angles (for a transversal)
    defined, 506
    concept development, 414, 416, 440, 443
corresponding parts
    defined, 506
    finding, 416
    of similar polygons, 416, 421–422, 483
cosine (cos), 461, 507
counterexamples
    defined, 33, 507
    concept development, 28–29, 33, 428,
        490, 492
counting numbers. *See* natural numbers
curve-fitting graphs, concept development,
    352–353, 354, 355, 361, 362, 363, 387,
    388–389, 409

D

data
    organization of. *See* In-Out tables
    representation of. *See* graphs and
        graphing
    spread of. *See* spread of data
    concept development, one-variable,
        190–191, 306, 307, 308, 309, 311,
        315, 316–317, 318, 319–320, 323,
        324–325, 326–329, 330, 331–332, 333
    concept development, two-variable,
        217–218, 219, 220, 221, 222–223,
        224–225, 229, 230, 304, 352–353,
        354, 355, 361, 371, 372, 387, 388–389
deductive reasoning, concept development,
    7, 51, 52, 70–71, 100–101, 184, 274–275,
    334–335, 349–350, 476
degrees
    calculators and, 462
    defined, 34, 507
    concept development, 27
dependent variables, 213, 507
diagonals
    defined, 507
    number of, 42, 70–71, 268
    and sum of angles of a polygon, 67
    concept development, 42, 70–71, 482
direct variation, concept development, 401,
    402, 407, 409, 458
discrete graphs
    as choice for representation, 210
    defined, 507
    concept development, 205–206, 210,
        211–212, 213
distributive property, concept
    development, 43, 44–45, 72, 187,
    195–196, 198–199, 200–201, 203, 228,
    239, 270, 271
division
    model for, 57
    negative integers and, 57
    in order of operations, 19

divisors
>   defined, 507
>   prime numbers and, 63

domain, 507

double-size polygons, 491

E

edges
>   of a cube, 272
>   defined, 507

elliptic geometry, 494

equally likely outcomes
>   area models and, 97, 98–99, 105
>   gambler's fallacy and, 88–89
>   sampling and, 90
>   sequences and, 145–146
>   concept development, 86, 94, 97, 98–99, 105, 106, 117, 145–146

equations
>   defined, 507
>   dependent variables and, 213, 507
>   equivalent. *See* equivalent equations
>   graphs and, 211–212, 357–358
>   independent variables and, 213, 510
>   for lines. *See* linear equations
>   order of operations and, 19
>   with proportions, simplifying, 456
>   for standard deviation, 327
>   solving. *See* solving equations
>   systems of. *See* systems of equations
>   concept development, 186–187, 215, 241–242, 243, 244, 245, 246, 247, 450
>   *See also* formulas; functions

equilateral triangles
>   defined, 507
>   concept development, 428, 437–438

equivalent equations
>   defined, 508
>   solving equations with, 257–258
>   concept development, 255–256, 257–258, 259

equivalent expressions
>   defined, 508
>   concept development, 12, 30–31, 43, 44–45, 160

even numbers, 107, 390

events, 508

expected value
>   defined, 508
>   fair games and, 114
>   number of outcomes and, 160
>   concept development, 109, 110, 113, 114, 115, 116, 117, 118, 119, 120, 121, 122, 123, 124, 125, 126, 127, 130, 131, 132, 132–133, 134, 136, 157, 158, 159, 160, 161, 163, 164, 167, 168, 170, 172

experimental estimate, 111–112

experimental probability. *See* observed probability

experimentation
>   correctness of mathematical models and, 402
>   In-Out tables and, 403, 404
>   process of, 305, 473
>   recording results of, 401

exponential functions, concept development, 202–203, 287

exponential graphs, concept development, 287, 356, 359

exponents
>   defined, 508
>   use in expressions, 18
>   *See also* square roots; squares (of a number)

expressions
>   algebraic. *See* algebraic expressions
>   defined, 508
>   equivalent. *See* equivalent expressions
>   meaning of, 18, 199
>   numeric. *See* numeric expressions
>   simplifying, concept development, 193, 252–253, 254, 255–256, 257–258

extensions of a problem, 9, 20

exterior angles
>   alternate, concept development, 440, 443, 444, 445
>   defined, 495, 508

linear functions
    adjustment to fit data, 229
    defined, 511
    the four representations for, 216, 233, 238
    general equation for, 238
    rate of change. *See* rate
    starting point, 238, 516
    technology to graph, 226–227, 229
    concept development, 211–212,
        217–218, 224–225, 229, 230, 234,
        236–237, 238, 239, 241–242, 243,
        244, 245, 246, 247, 250–251,
        260–261, 354, 355
linear graphs, concept development,
    211–212, 217–218, 224–225, 234,
    236–237, 238, 239, 241–242, 243, 244,
    245, 246, 247, 250–251, 285, 286, 357
line of best fit
    defined, 511
    In-Out table from, 217
    sketching, 217, 221
    concept development, 217–218, 220,
        221, 222–223, 224–225, 229, 230
line segments, 511
logic and reasoning
    concept development, 100–101, 276,
        300–301, 310, 314, 379, 380
    See also deductive reasoning; inductive
        reasoning
lower limit, 30
lower quartile, 376–377, 511

M

market analysis, 352–353
mathematical models
    area models. *See* area models
    correctness of, 402
    defined, 511
    diagrams refined into, 402
    purpose of, 402
mean
    defined, 511
    frequency bar graph and, 142
    range as measure of spread of, 323

and standard deviation, 326, 328, 329
    symbol for, 326
    concept development, 87, 113, 114,
        116, 141, 142–143, 190–191, 318,
        319–320, 330, 331–332, 333, 336,
        337–338, 341, 342–343, 361, 381,
        382, 383
    See also measures of center
measurement of angles. *See* angles
measurement variation
    defined, 511
    normal distribution and, 327–328
    concept development, 36, 37, 38, 302,
        303, 306, 307, 308–309, 311,
        316–317, 318, 319–320, 336,
        337–338, 341, 342–343, 430, 459
measures of center
    average. *See* average
    mean. *See* mean
    median. *See* median
    mode. *See* mode
    concept development, 117, 142–143,
        190–191, 306, 307, 308, 309, 313,
        315, 318, 319–320
    See also expected value
median
    box plots and, 376–377
    defined, 511
    frequency bar graph and, 143
    concept development, 87, 142–143,
        190–191
mode
    defined, 511
    concept development, 87
models. *See* mathematical models
multiplication
    with negative numbers, 22–23
    order of operations for, 19
    properties for. *See* real number
        properties
multiplicative identity, concept
    development, 21–23, 26, 56, 57

S

samples, concept development, 87, 88–89, 90, 92–93, 103, 104, 109, 111–112, 116, 121, 124, 125, 126
sample standard deviation, defined, 515
scale factor
    concept development, 414, 415, 416, 434–435, 465, 478, 479
    See also transformations
scaling
    of axes, 208–209, 210
    of drawings and models, 479, 481
    concept development, 208–209, 210, 217–218
scatterplots, concept development, 217–218, 219, 220, 361
self-assessment, 9
sequences
    consecutive, 28
    defined, 515
    equally likely outcomes and, 145–146
    high-low, 282–283
    nth term of, 49
    as pattern, 4
    results reported as, 145–146
    technology used to generate, 392–393
    terms of, 4, 517
    concept development, 4, 25, 49, 390–391, 392–393
sigma notation. *See* summation notation
sigma symbol, 30
similar polygons
    corresponding parts and, 416, 421–422, 483
    defined, 515
    diagrams and, 412–413
    equations and, 425
    triangles. *See* similar triangles
    concept development, 412–413, 414, 415, 416, 421–422, 423–424, 428, 429–430, 431, 432, 433, 437–438, 443, 447, 448, 449, 450, 451, 459, 465, 468, 482, 483, 484, 490, 492, 495–497, 546

similar triangles
    angle measures and, 433, 434, 485, 487
    characteristics of, 428, 431
    height measurement using, 449, 456
    overlapping, 432, 443
    ratio of sides and, 421
    sufficient conditions for, 485
    trigonometric ratios and, 460, 463
    See also triangles; trigonometry
simulation
    defined, 515
    with graphing calculator, 121, 126
    concept development, 109, 111–112, 116, 121, 124, 125, 126, 127, 136
sine (sin), 460–461, 515
slope
    defined, 515
    concept development, 211–212, 217–218, 220, 221, 222–223, 224–225, 229, 230, 234, 236–237, 238, 239, 285, 286, 291
    See also rate
solutions
    absolute certainty, defined, 7
    defined, 515
    of equivalent equations, 255
    simpler related problem, 128
    working backward, 74–77
    write-up of, 9
    See also counterexamples; generalizing
solving equations, concept development, 186–187, 211–212, 215, 226–227, 228, 252–253, 254, 255–256, 257–258, 259, 260–261, 421–422, 423–424, 425, 426, 429–430, 432, 434–435, 445, 449, 450, 469–470
spiralaterals, 453–455
spread of data
    range as measure of. *See* range
    variance, 327, 518
    concept development, 324–325, 381
    See also standard deviation

terms (of an expression), 517

terms (of a sequence), 4, 517

theoretical probability

    area models. *See* area models

    as choice to find probability, 92–93

    defined, 517

    concept development, 87, 92–93, 95, 97, 105, 106, 109, 111–112, 117, 118, 124, 125, 127, 130, 151, 152, 153, 158

    See also probability

transformations

    concept development, 13, 412–413

    See also reflections; rotations; scale factor; translations

translations, concept development, 412–413, 453–455

transversals

    defined, 517

    and parallel lines, 440

    and parallel postulate, 493

    concept development, 440, 443, 444, 445, 451

trapezoids, 491, 517

triangle inequality principle

    defined, 518

    concept development, 431, 436, 441–442

triangles

    angle sum property. *See* triangle sum property

    congruent, 433, 487

    defined, 517

    determination of, 487–488

    double-size, 491

    equilateral. *See* equilateral triangles

    inequality principle. *See* triangle inequality principle

    isosceles. *See* isosceles triangles

    right. *See* right triangles

    rigidity of, 484

    similar. *See* similar triangles

    sum of angles of, 36, 67, 444

    concept development, 36, 42, 416, 421–422, 425, 428, 429–430, 431, 447, 465, 491

    See also polygons; trigonometry

triangle sum property

    proof of, 444

    concept development, 36, 67, 428, 433, 436, 437–438, 439, 444, 451

trigonometric functions

    defined, 518

    concept development, 463, 464

trigonometry

    calculators for, 462

    cosine, 461, 507

    defined, 460

    similarity and, 460, 463

    sine, 460–461, 515

    table of values for, 463

    tangent, 461, 517

    concept development, 459, 460–462, 463, 464, 465, 466, 467, 469–470, 498, 499, 500, 501

U

unique (solution), 63, 518

upper limit, 30

upper quartile, 376–377

    defined, 518

V

variables

    defined, 518

    experiments to explore, guidelines for, 399–400

    concept development, 10, 44–45, 186–187, 193, 194, 195–196, 197, 198–199, 200–201, 241–242, 243, 269–270, 271, 303, 342–343, 347, 398–400, 401, 402, 407, 408, 409, 410, 458, 472, 473

variance

    defined, 518

    equation for standard deviation and, 327

vertex (vertices)

    of an angle, 34

    of a polygon, 35

    defined, 518

INDEX OF ACTIVITY TITLES

PHOTOGRAPHIC CREDITS

Front Cover Photos

Berkeley High School: Nader Al-Ammari, Mayra Cuevas, Antonio Lozano, Keahara Monroe, Sonia Mena, Hien Nguyen, Alejandra Pedroza, Jessica Quindel, Rodrick Rogers, Sean Scott, Suraj Verma; Harry S. Truman High School: Robert Muniz

Front Cover and Unit Opener Photography

Truman High School photos: Arlene DeSimone; Berkeley High School photos: Cheryl Fenton; iStockphoto

Patterns

1 (from top left, clockwise) Hong Kong Chen/iStockPhoto; Nan Moore/iStockPhoto; Harry S. Truman High School, Arlene DeSimone; Berkeley High School, Cheryl Fenton; Brendan Hunter/iStockPhoto; Berkeley High School, Cheryl Fenton; Berkeley High School, Cheryl Fenton; **3** Berkeley High School, Cheryl Fenton; **13** Little Blue Wolf Productions/Corbis; **15** Lincoln High School, Lori Green; **27** San Lorenzo Valley High School, Kim Gough; **28** Mark Karrass/Corbis; **32** Simon Jarratt/Corbis; **39** Lincoln High School, Lori Green; **46** Berkeley High School, Cheryl Fenton; **49** Comstock/Corbis; **51** KCP; **54** KCP; **55** Thinkstock/Corbis; **56** KCP; **58** KCP; **69** Cone 6 Productions/Brand X/Corbis; **74–77** Irina Ovchinnikova/iStockPhoto

The Game of Pig

79 (from top left, clockwise) iStockPhoto; Simone van den Berg/iStockPhoto; Harry S. Truman High School, Arlene DeSimone; Berkeley High School, Cheryl Fenton; Brent Melton/iStockPhoto; Berkeley High School, Cheryl Fenton; Berkeley High School, Cheryl Fenton; **81** Foothill High School, Cheryl Dozier;

365 Berkeley High School, Cheryl Fenton; **366** Berkeley High School, Cheryl Fenton; **370** Photoedit; **372** Tetra Images/Corbis; **383** KCP; **390** Ivan Hunter/Getty

Shadows

395 (from top left, clockwise) Meredith Mullins/iStockPhoto; Duncan Walker/iStockPhoto; Harry S. Truman High School, Arlene DeSimone; Berkeley High School, Cheryl Fenton; Dave Perkins/iStockPhoto; Berkeley High School, Cheryl Fenton; Berkeley High School, Cheryl Fenton; **397** Santa Cruz High School, Lynne Alper; **401** Muzium Negara; **402** VisionsofAmerica/Joe Sohm/Getty; **405** Stock Montage; **411** Santa Maria High School, Mike Bryant; **427** Berkeley High School, Cheryl Fenton; **428 (top)** Carlos E. Santa Maria/Shutterstock; **428 (bottom)** Adrian Matthiassen/Shutterstock; **436** KCP; **446** Santa Maria High School, Chris Paulus; **457** Berkeley High School, Cheryl Fenton; **468** Berkeley High School, Cheryl Fenton; **469** Berkeley High School, Cheryl Fenton; **472** KCP; **479** Peter Barrett/iStockPhoto; **480** KCP; **483** InMagine Photo; **484** Library of Congress; **485 (left)** Doug Schneider/iStockPhoto; **485 (right)** Foodcollection/Getty; **489** Jupiter Images; **493** Getty Images; **498** Michele Constantini/Getty

Acknowledgements

My name gets to grace the cover of this book, but this is well and truly a team effort. Firstly, enormous thanks and big love to Melissa Clark (foodenvy.com.au), my partner in crime for recipe development and for so beautifully styling our dishes for our photo shoots. Mel, your work is fabulous and food delicious, I couldn't have done this without you.

I'm lucky enough to have worked with two amazing photographers on our recipe shoots—Nicky Ryan (nickyryan.com) and Nicholas Wilson (nicholaswilson.com.au). Guys your attention to detail and artistic flare brings our food to life—thank you for your expertise and patience on shoot days! Thanks too to the third and newest member in my recipe team, Jennifer Jenner (84thand3rd.com). JJ not only develops a recipe to my specifications, she styles and does the photography. JJ you are such a talent and I'm so lucky to have you join my team.

In addition to the recipe shoots we had a day to shoot the cover along with a couple of additional lifestyle shots for the pages of the book. My thanks to Vivien Valk art director, Tracy Rutherford who styled the set, Simone Cozens for hair and makeup, and photographer Steve Brown. You made my day in front of the camera very easy!

At Murdoch Books I've been exceptionally fortunate to have scored a team of professionals who understood my ideas and worked their magic to bring them all to life. To my publisher Corinne Roberts and editor Emma Hutchinson, thank you for having faith in my concept and helping to chisel my tome of a manuscript into a concise and polished book. You have both been fantastic to work with and I truly appreciate your help and flexibility in accommodating my schedules. I have to also thank all those at Murdoch Books who have worked behind the scenes to pull this book together from the designers to those in sales and marketing. Without you this book would never have come to life.

This is also my chance to say a huge thank you to my business manager and right-hand woman, Sasha Kahan McSweeney. Sasha, you're a joy to work with, you keep me on track and together I have no doubt we will continue to grow.

Finally, to my husband Joel, who puts up with my irregular work hours and my distracted mind while working. Thank you babe for your bottomless support and faith in my abilities, your irreplaceable help in running my business and for making our life together an awesome, joyful ride!

Published in 2017 by Murdoch Books, an imprint of Allen & Unwin

Murdoch Books Australia
83 Alexander Street
Crows Nest NSW 2065
Phone: +61 (0) 2 8425 0100
Fax: +61 (0) 2 9906 2218
murdochbooks.com.au
info@murdochbooks.com.au

Murdoch Books UK
Ormond House
26–27 Boswell Street
London WC1N 3JZ
Phone: +44 (0) 20 8785 5995
murdochbooks.co.uk
info@murdochbooks.co.uk

For Corporate Orders & Custom Publishing, contact our Business
Development Team at salesenquiries@murdochbooks.com.au.

Publisher: Corinne Roberts
Editor: Emma Hutchinson
Cover designer: Vivien Valk
Page designer: Nicky Kukulka
Illustrator: Nicky Hodgson, pages 12,14, 15, 18, 31, 32, 37, 45, 52 and 54
Additional illustrations: iStock, pages 22, 27, 33 and 39; Shutterstock, page 35
Photographers: Steve Brown, Jennifer Jenner, Nicky Ryan and
    Nicholas Wilson
Principal food stylist and recipe developer: Melissa Clark
Additional food styling and recipe development: Jennifer Jenner
Location shoot stylist: Tracy Rutherford
Production Manager: Rachel Walsh

A cataloguing-in-publication entry is available from the catalogue of the
National Library of Australia at nla.gov.au.

ISBN 978 1 74336 848 0 Australia
ISBN 978 1 74336 850 3 UK

A catalogue record for this book is available from the British Library.

Colour reproduction by Splitting Image Colour Studio Pty Ltd,
    Clayton, Victoria

Printed by 1010 Printing International Limited, China

DISCLAIMER: The purchaser of this book understands
that the information contained within is not intended
to replace medical advice. It is understood that you
will seek full medical clearance by a licensed physician
before making any changes mentioned in this book.
The author and publisher claim no responsibility to
any person or entity for any liability, loss, or damage
caused or alleged to be caused directly or indirectly as
a result of the use, application or interpretation of the
material in this book.

IMPORTANT: Those who might be at risk from the
effects of salmonella poisoning (the elderly, pregnant
women, young children and those suffering from
immune deficiency diseases) should consult their
doctor with any concerns about eating raw eggs.

OVEN GUIDE: You may find cooking times
vary depending on the oven you are using. For
conventional ovens, as a general rule, set the oven
temperature to 20°C (70°F) higher than indicated
in the recipe.

MEASURES GUIDE: We have used 20 ml (4 teaspoon)
tablespoon measures. If you are using a 15 ml
(3 teaspoon) tablespoon add an extra teaspoon of
the ingredient for each tablespoon specified.